REVISE EDEXCEL GCSE
Mathematics
Specification A Linear

REVISION GUIDE

Foundation

Series Director: Keith Pledger

Series Editor: Graham Cumming

Authors: Harry Smith, Gwenllian Burns, Jean Linsky

- -

A note from the publisher

For the full range of Pearson revision titles across GCSE, BTEC and AS Level visit:
www.pearsonschools.co.uk/revise

ALWAYS LEARNING

PEARSON

Contents

A small bit of small print

A grade allocated to a question represents the highest grade covered by that question. Sub-parts of the question may cover lower grade material.

The grade range of a topic represents the usual grade range that the topic is assessed at. The topic may form part of a higher grade question if tested within the context of another topic.

Questions in this book are targeted at the grades indicated.

C
D
E
F
G

Place value

The value of each digit in a number depends on its position. Digits which are further to the left are worth more. You can use a place value diagram to help you read and write numbers.

Worked example

grade **G**

(a) Write the number 6804 in words.
Six thousand, eight hundred and four

(b) Write the number **nine thousand and fifty-eight** in figures.
9058

(c) Write the number **fifteen thousand, four hundred and twenty-five** in figures.
15 425

(d) Write down the value of the 6 in 28 674
600

	Ten thousands	Thousands	Hundreds	Tens	Units
(a)		6	8	0	4
(b)		9	0	5	8
(c)	1	5	4	2	5
(d)	2	8	⑥	7	4

You do not need to write 'no tens'.

Use a zero to fill any empty spaces.

600 or 6 hundreds

Worked example

grade **G**

Write these numbers in order of size, <u>starting with the smallest</u>.

2908	2950	5011	925	10 430

925				
925				10 430
925			5011	10 430
925	2908	2950	5011	10 430

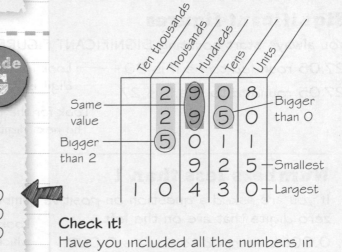

	Ten thousands	Thousands	Hundreds	Tens	Units
Same value		2	9	0	8
		2	9	5	0
Bigger than 2		5	0	1	1
			9	2	5
	1	0	4	3	0

Bigger than 0

Smallest

Largest

Check it!
Have you included all the numbers in your final answer? ✓

Rounding

You may be asked to write a number to the nearest 10, 100 or 1000.

5849 to the nearest 1000 is 6000.   5849 is nearer 6000 than 5000.
5849 to the nearest 100 is 5800.   5849 is nearer 5800 than 5900.

For a reminder about rounding have a look at page 2.

Now try this

edexcel

54 327 people watched a concert.
(a) (i) Write 54 327 in words. **(1 mark)**
  (ii) Write down the value of the 3 in the number 54 327 **(1 mark)**
  (iii) Write the number 54 327 to the nearest thousand. **(1 mark)**

(b) Write these numbers in order of size. Start with the smallest number.

grade **G**

| 35 687 | 3091 | 29 999 |
| 104 152 | 3100 |

(2 marks)

C
D
E
F
G

Rounding numbers

To ROUND a number, you look at the next digit to the right on a place value diagram.

5 or more → round up, less than 5 → round down

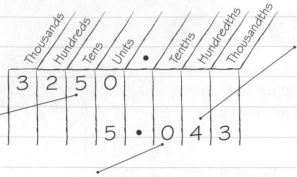

To round to the nearest 100, you look at the digit in the tens column.
It is a 5, so round up. 3250 rounded to the nearest 100 is 3300.

To round to the nearest whole number look at the digit in the tenths column.
It is a 0, so round down.
5.043 rounded to the nearest whole number is 5.

To round to 1 decimal place (1 d.p.), you look at the digit in the second decimal place.
It is a 4, so round down.
5.043 rounded to 1 d.p. is 5.0
You NEED to write the 0 to show that you have rounded to 1 d.p.

Significant figures

You always start counting SIGNIFICANT FIGURES from the left.

27.05 rounded to 1 s.f. is 30 ⟶ Look for the digit furthest to the left, which is 2. The next digit is 7, so round up to give an answer of 30

27.05 rounded to 2 s.f. is 27 ⟵ Look for the two digits furthest to the left, which are 2 and 7. The next digit is 0, so round down to give an answer of 27

Numbers less than 1

If you are asked a question on positive numbers less than 1, remember NOT to count zero digits that are on the left.

0.0085 rounded to 1 s.f. is 0.009 ⟶ Look for the digit which is furthest to the left and which is NOT a zero. This digit is 8. The next digit is 5 so round up to give an answer of 0.009

Worked example

grade D

Round 6.9083 correct to
(a) 1 significant figure
(b) 2 significant figures
(c) 3 significant figures

(a) 7 (1 s.f.)   (b) 6.9 (2 s.f.)   (c) 6.91 (3 s.f.)

(a) The first digit is 6.
The next digit is 9 so round up.
(b) The second digit is 9.
The next digit is 0 so round down.
(c) The third digit is 0.
The next digit is 8 so round up.

Now try this

edexcel ⠿

(a) In one week 9849 people visited a museum.
Round the number 9849 to the nearest hundred. **(1 mark)**

(b) A café had 23 578 customers last year.
Round the number 23 578 to 3 significant figures. **(1 mark)**

grade D

Adding and subtracting

You need to be able to add and subtract numbers without a calculator.

Mental methods

Try these methods for adding and subtracting quickly in your head.

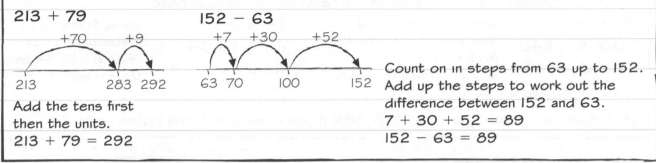

213 + 79

+70   +9

213   283 292

Add the tens first
then the units.
213 + 79 = 292

152 − 63

+7 +30   +52

63 70   100   152

Count on in steps from 63 up to 152.
Add up the steps to work out the
difference between 152 and 63.
7 + 30 + 52 = 89
152 − 63 = 89

Worked example   grade **G**

Work out 285 + 56 + 1091

```
   285
    56
+ 1091
------
  1432
   2 1
```

1. **Always** add the units column first.
   5 + 6 + 1 = 12. Write down the 2 and carry the
   1 over to the tens column.
2. Add the tens column. 8 + 5 + 9 + 1 = 23
   Make sure you include any numbers you carried
   over. Write down the 3 and carry the **2** over to
   the hundreds column.
3. Add the hundreds column. 2 + 0 + **2** = 4 Write
   down 4.
4. There is only one digit in the thousands column.
   Write this in your answer.

Worked example   grade **G**

Work out 418 − 62

```
  ³4̷¹18
 −  62
-----
   356
```

1. **Always** subtract the units column first.
   8 − 2 = 6
   Remember it is (top number) − (bottom number)
2. Look at the tens column. 1 is smaller than 6 so
   you have to exchange 1 hundred for 10 tens.
   Change 4 hundreds into 3 hundreds and 10 tens.
3. Now you can subtract the tens column. 11 − 6 = 5
4. There is nothing to subtract in the hundreds
   column so write 3 in your answer.

Now try this   grade **F**

Subtract the number of people
that got off and add the number
of people that got on.

1. 33 people were on a bus.
   At the first stop, 19 people got off and
   15 people got on.
   At the second stop, 17 people got off and
   26 people got on.
   How many people are now on the bus?

   (3 marks)

2. Nathan goes shopping. He buys
   2 books costing £4.40 **each**
   1 magazine costing £2.90
   He pays with a £20 note.
   Work out how much change he should get.

   (4 marks)

grade **F**

edexcel

C
D
E
F
G

Multiplying and dividing

You need to be able to multiply and divide numbers without a calculator.
For a reminder about multiplying and dividing by 10, 100 and 1000 have a look at page 62.

Mental methods

Try these methods for multiplying and dividing quickly in your head.

37×8

$30 \times 8 = 240$

$7 \times 8 = 56$

$37 \times 8 = 296$

Split 37 into 30 and 7 then multiply each by 8. Add each separate answer to get the total.

$54 \div 6$

$6 \times \square = 54$

The answer is 9.

Try to find a multiplication fact using 6 with 54 as the answer.

Multiplying and dividing are MUCH EASIER if you know your times tables up to 10×10.

Worked example

grade G

Work out
(a) 49×3

$$\begin{array}{r} 49 \\ \times 3 \\ \hline 147 \\ {}^2 \end{array}$$

(b) 36×24

$$\begin{array}{r} 36 \\ \times 24 \\ \hline 144 \\ {}_2 \\ 7{,}20 \\ \hline 864 \end{array}$$

Always multiply from right to left.
1. $9 \times 3 = 27$. Write down 7 and carry over **2** (2 tens).
2. $4 \times 3 = 12$. Add on the carry-over. $12 + \mathbf{2} = 14$. Write down 14.

EXAM ALERT!

Half of students did not get any of the 3 marks available for part (b). You have to show your working to get them all.
1. Work out 36×4. (Answer = 144)
2. Work out 36×20. Write down **0** and then work out 36×2. (Answer = 720)
3. Add the separate answers. $(144 + 720 = 864)$

This was a real exam question that caught students out – **be prepared!**   Results**Plus**

Worked example

grade G

Work out $288 \div 9$

$$\begin{array}{r} 32 \\ 9{\overline{)}}288 \\ -27 \\ \hline 18 \\ -18 \\ \hline 0 \end{array}$$

You can use a method called long division for written division.
1. Does 9 divide into 2? No.
2. Does 9 divide into 28? Yes.
   $9 \times 3 = 27$ so 9 divides into 28 three times with remainder 1.
3. Does 9 divide into 18? Yes.
   $9 \times 2 = 18$ so 9 divides into 18 two times with no remainder.
Using short division the calculation would look like this:

$$\begin{array}{r} 32 \\ 9{\overline{)}}28^18 \end{array}$$

Now try this

You will need to multiply for part (a), and divide for part (b).

grade E

edexcel

A box of chocolate bars contains 46 bars.
(a) Work out the total number of bars in 35 boxes. **(3 marks)**

368 bars of chocolate are shared equally between 8 groups.
(b) How many bars does each group receive? **(3 marks)**

Decimals and place value

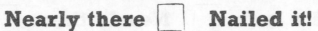

C
D
E
F
G

You can use a place value diagram to help you understand and compare decimal numbers. Remember that decimal numbers with more digits are not necessarily bigger. Try writing extra Os so that all the numbers have the same number of decimal places.

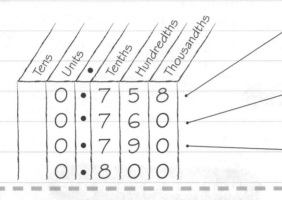

The value of the 5 in this number is 5 hundredths.

0.76 is the same as 0.760
0.76 is bigger than 0.758
6 hundredths is bigger than 5 hundredths.

0.79 is smaller than 0.8 because the digit in the tenths place is smaller.

Worked example   grade F

Write the following numbers in order of size. <u>Start with the smallest number.</u>

0.32   0.315   0.3   0.39   0.379

~~0.32~~ ~~0.315~~ ~~0.3~~ ~~0.39~~ ~~0.379~~

0.3   0.315

0.3   0.315   0.32

0.3   0.315   0.32   0.379   0.39

EXAM ALERT!

About 60% of students got this wrong. All the numbers have the same tenths digit so look at the hundredths digit first. 0.3 is the same as 0.30 so this is the smallest number.

Check it!
Have you included all the numbers in your final answer? ✓

This was a real exam question that caught students out – **be prepared!**   ResultsPlus

---

Worked example   grade D

Using the information that
58 × 71 = 4118
write down the value of
(a) 58 × 0.71
41.18
(b) 5800 × 7.1
41180

(a) 71 has been divided by 100 and 58 hasn't been changed. So the answer needs to be divided by 100: 4118 ÷ 100 = 41.18
(b) 58 has been multiplied by 100 and 71 has been divided by 10.

 is the same as ─×10─→

The answer needs to be multiplied by 10:
4118 × 10 = 41180
For a reminder about multiplying and dividing by 10, 100 and 1000 see page 62.

Now try this   grade F

edexcel

1. (a) Write down the value of the 9 in the number 3.091   **(1 mark)**
   (b) Write these numbers in order of size. Start with the smallest number.
      (i) 0.56, 0.067, 0.6, 0.65, 0.605
      (ii) 0.72, 0.7, 0.072, 0.07, 0.702   **(2 marks)**

2. Using the information that
   32 × 61 = 1952
   write down the value of   grade D
   (a) 3.2 × 61   **(1 mark)**
   (b) 0.32 × 6100   **(1 mark)**

C
D
E
F
G

Operations on decimals

1 Adding and subtracting

To add or subtract decimal numbers:
1. Line up digits with the same place value.
2. Line up the decimal points.
3. Write a decimal point in your answer.

See page 3 for a reminder about adding and subtracting.

> Write in 0s so that both numbers have the same number of decimal places.

Worked example grade F

Work out
(a) $0.75 + 1.6$

```
  0.75
+ 1.60
------
  2.35
```

(b) $3.5 - 0.21$

```
  3.⁴5̷0
- 0.21
------
  3.29
```

2 Multiplying

To multiply decimal numbers:
1. Ignore the decimal points and just multiply the numbers.
2. Count the number of decimal places in the calculation.
3. Put this number of decimal places in the answer.

(a) You can use estimation to check that the decimal point is in the correct place.
$8.69 \times 12 \approx 9 \times 12 = 108$ $108 \approx 104$ ✓

(b) 8.5×0.04 has 3 decimal places in total

$$.3\overset{\frown}{4}\overset{\frown}{0} = 0.34$$

Write a 0 before the decimal point and simplify your answer.

Worked example grade D

Work out   (b) 8.5×0.04
(a) 8.69×12 grade F

```
    869
×    12
------
  1738
+ 8690
------
 10428
```

$8.69 \times 12 = 104.28$

```
   85
×   4
-----
  340
```

$8.5 \times 0.04 = 0.34$

3 Dividing

To divide by a decimal number:
1. Multiply both numbers by 10, 100 or 1000 to make the second number a whole number.
2. Divide by the whole number.

> Multiply 40.6 and 1.4 by 10.
> If you multiply both numbers in a division by the same amount, the answer stays the same.

Worked example grade D

(a) $55.8 \div 3$

```
    1 8.6
3)5²5.¹8
```

(b) $40.6 \div 1.4$

$40.6 \div 1.4 = 406 \div 14$

```
    29
14)406
  -28
  ---
   126
  -126
  ----
     0
```

$40.6 \div 1.4 = 29$

Now try this

edexcel

A ream of paper costs £2.70
(a) Work out the cost of 43 reams of paper.
(2 marks)

Rob has £78
(b) (i) Work out the greatest number of reams of paper he can buy. **(2 marks)**
(ii) How much money will he have over? **(2 marks)**

grade D

Estimating answers

You can estimate the answer to a calculation by rounding each number to 1 SIGNIFICANT FIGURE and then doing the calculation.

This is useful for checking your answers.

$4.32 \times 18.09 \approx 4 \times 20 = 80$

$\approx$ means 'is approximately equal to'.

Significant figures

To round a number to 1 significant figure (s.f.), look at the digit to the right of the first non-zero digit.

4.32 rounded to 1 s.f. is 4
The first non-zero digit is 4. The next digit is 3 so round down.

For a reminder about significant figures look at page 2.

Dividing by a number less than 1

Some calculations will involve division by a number less than 1.

If you multiply both numbers in a division by the same amount the answer stays the same. To divide by 0.5, start by multiplying the top and bottom by 10.

$\dfrac{6.29 \times 340}{0.461} \approx \dfrac{6 \times 300}{0.5} = \dfrac{1800}{0.5} = 3600$

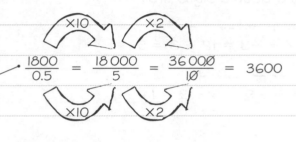

$\dfrac{1800}{0.5} = \dfrac{18\,000}{5} = \dfrac{36\,000}{10} = 3600$

For more on dividing by a decimal look back at page 6.

Worked example　grade C

Work out an estimate for the value of

(a) $\dfrac{7.82 \times 620}{0.525}$

$\approx \dfrac{8 \times 600}{0.5} = \dfrac{4800}{0.5} = \dfrac{48\,000}{5} = 9600$

(b) $\dfrac{6.8 \times 191}{0.051}$

$\approx \dfrac{7 \times 200}{0.05} = \dfrac{1400}{0.05} = \dfrac{140\,000}{5} = 28\,000$

EXAM ALERT!

Students found part (b) particularly difficult.

Do **not** round the numbers to the nearest whole number. You should round each number to **1 significant figure**.

It is important to show each of the rounded numbers **and** each step of your calculation.

For division by 0.05, you can multiply the top and bottom by 100:

$\dfrac{1400}{0.05} = \dfrac{140\,000}{5} = \dfrac{280\,000}{10} = 28\,000$

This was a real exam question that caught students out – **be prepared!**　Results Plus

Now try this　grade C　edexcel ▦

1. Work out an estimate for the value of

$\dfrac{637}{3.2 \times 9.8}$　**(3 marks)**

2. Work out an estimate for

$\dfrac{410 \times 6.9}{0.23}$　**(3 marks)**　grade C

C
D
E
F
G

Negative numbers

Numbers smaller than 0 are called NEGATIVE numbers.

⟨NEGATIVE NUMBERS | POSITIVE NUMBERS⟩

−5 −4 −3 −2 −1 0 1 2 3 4 5

0 is neither positive nor negative.

You can use a number line to write numbers in order of size. The numbers get bigger as you move to the right.

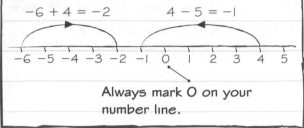

Number lines

You can use number lines to help when adding and subtracting.

$-6 + 4 = -2$ $4 - 5 = -1$

−6 −5 −4 −3 −2 −1 0 1 2 3 4 5

Always mark 0 on your number line.

Adding and subtracting

To add or subtract a negative number, change the double signs first.

$+ - \rightarrow -$

$12 + -3 = 12 - 3 = 9$

$- - \rightarrow +$

$5 - -9 = 5 + 9 = 14$

Multiplying and dividing

When multiplying and dividing, use these rules to decide whether the answer will be positive or negative.

Numbers have the SAME sign
→ answer is POSITIVE

$-3 \times -7 = 21$

Numbers have DIFFERENT signs
→ answer is NEGATIVE

$+80 \div -10 = -8$

Worked example

grade F

One day the temperature at midnight was −7 °C, the temperature at 10 am was −1 °C and the temperature at midday was 3 °C.

Jenny says that, on this day, the temperature at 10 am is halfway between the temperatures at midnight and at midday.

Is Jenny correct?

You must give a reason for your answer.

No. −2 °C would be halfway between −7 °C and 3 °C.

EXAM ALERT!

Fewer than half of students got this question right. Read the question carefully and draw a number line to help you.

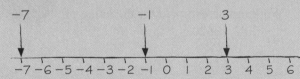

−7   −1   3

−7 −6 −5 −4 −3 −2 −1 0 1 2 3 4 5 6

You could also give these reasons: 'The difference between −7 and −1 is 6 and the difference between −1 and 3 is only 4' or '−1 is closer to 3 than to −7.'

Now try this

edexcel

(a) Work out   (i) $3 - 11$ (ii) -3×-5

(2 marks)

(b) Work out $\dfrac{-8 \times -3}{-6}$

(2 marks)

(c) Write these numbers in order of size. Start with the smallest number.

8   −5   3   −1   0

(2 marks)

grade F

Squares, cubes and roots

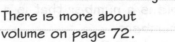

C
D
E
F
G

Squares and square roots

When a number is multiplied by itself the answer is a SQUARE NUMBER. You can write square numbers using index notation.

Multiplication	Index notation	Square number
2 × 2	2^2	4
5 × 5	5^2	25
9 × 9	9^2	81
13 × 13	13^2	169

Square numbers are the areas of squares with whole number side lengths.

25 cm² | 5 cm
5 cm

There is more about area on page 65.

SQUARE ROOTS are the opposite of squares. You use the symbol $\sqrt{}$ to represent a square root.

$\sqrt{4} = 2$ $\sqrt{25} = 5$ $\sqrt{81} = 9$

You need to be able to REMEMBER the square numbers up to 15 × 15 and the corresponding square roots.

Cubes and cube roots

When a number is multiplied by itself then multiplied by itself again, the answer is a CUBE NUMBER. You can write cube numbers using index notation.

Multiplication	Index notation	Cube number
2 × 2 × 2	2^3	8
3 × 3 × 3	3^3	27
5 × 5 × 5	5^3	125
10 × 10 × 10	10^3	1000

Cube numbers are the volumes of cubes with whole number side lengths.

27 cm³ | 3 cm
3 cm | 3 cm

There is more about volume on page 72.

CUBE ROOTS are the opposite of cubes. You use the symbol $\sqrt[3]{}$ to represent a cube root.

$\sqrt[3]{8} = 2$ $\sqrt[3]{27} = 3$ $\sqrt[3]{1000} = 10$

You need to be able to REMEMBER the cubes of 2, 3, 4, 5 and 10, and the corresponding cube roots.

EXAM ALERT!

Only one-third of students got full marks on this question.

It's much easier to answer questions like this if you can **recognise** all the square numbers up to 15 × 15. To show that Scott is wrong you need to give an example of two square numbers which add together to make an odd number.

If you use more than one example, make sure you say which example shows that Scott is wrong.

This was a real exam question that caught students out – **be prepared!**   Results**Plus**

Worked example

grade
D

Scott says, 'If you add two different square numbers, you will always get an even number.' Show that Scott is **wrong**.

$2^2 = 4$
$3^2 = 9$
$4 + 9 = 13$
13 is an odd number

'Difference' means your example will be a square number take away another square number.

Now try this

edexcel :::

grade
F

1. (a) Write down
   (i) the cube root of 1000
   (ii) the square root of 4   **(2 marks)**
   (b) Work out (i) 7^2 (ii) 1^3 (iii) $\sqrt{144}$ (iv) $\sqrt[3]{64}$
      (4 marks)

2. Keri says, 'If you find the difference of two different square numbers, you will always get an odd number.'
   Keri is wrong. Explain why.   **(2 marks)**

grade
D

Had a look ☐ Nearly there ☐ Nailed it! ☐

C D E F G

Factors, multiples and primes

Factors and multiples

The FACTORS of a number are any whole numbers that divide into it exactly.

1 and the number itself are both factors of any number.
The factors of 12 are 1, 2, 3, 4, 6 and 12.

Factors come in pairs. Each pair is a multiplication fact with the number as its answer.
The factor pairs of 12 are 1 × 12, 2 × 6 and 3 × 4.

A common factor is a number that is a factor of two or more numbers.
2 is a common factor of 6 and 12.

The MULTIPLES of a number are all the numbers in its times table.
The multiples of 7 are 7, 14, 21, 28, 35, ...

A common multiple is a number that is a multiple of two or more numbers.
12 is a common multiple of 6 and 4.

Primes

A PRIME NUMBER has exactly two factors. It can only be divided by 1 and by itself.

The first ten prime numbers are
2, 3, 5, 7, 11, 13, 17, 19, 23, 29.
1 is not a prime number. It has only 1 factor.

Factor trees

You can use a factor tree to find prime factors.
1. Choose a factor pair of the number.
2. Circle the prime factors as you go along.
3. Continue until every branch ends with a prime number.
4. At the end write down ALL the circled numbers, putting in multiplication signs.

Worked example

16 3 20
8 17 6 12

Using the numbers in the cloud write down

(a) a prime number
17

(b) a multiple of 5
20

grade G

(c) the factors of 24 which have a sum of 15
12 and 3

grade F

(a) 3 is also a prime number.
(b) 3, 6, 8 and 12 are all factors of 24.
    Only 12 and 3 have a sum of 15.

Worked example

grade C

Write 90 as a product of prime factors.

```
        90
       /  \
      30   ③
     /  \
    10   ③
   /  \
  ⑤   ②
```

90 = 5 × 2 × 3 × 3

5 × 2 × 3 × 3 is called a **product of prime factors.** You can also write
5 × 2 × 3²

Now try this

grade G

1. | 3 | 4 | 5 | 6 | 8 | 12 | 16 | 27 | 30 |

   Using only the numbers in the box, write down
   (a) all the multiples of 6
   (b) all the prime numbers
   (c) two factors of 24 that add up to 20 **(3 marks)**

grade F

2. Write 168 as a product of its prime factors.
   (3 marks)

grade C

HCF and LCM

The HIGHEST COMMON FACTOR (HCF) of two numbers is the highest number that is a factor of both numbers.

The LOWEST COMMON MULTIPLE (LCM) of two numbers is the lowest number that is a multiple of both numbers.

For a reminder about prime factors and drawing a factor tree, have a look back at page 10. Don't forget to write out the product of primes at the end.

Check it!

$2 \times 2 \times 3 \times 3 \times 3 = 108$ ✓

To find the HCF circle all the prime numbers which are **common** to both products of prime factors. 2 appears twice in both products so you have to circle it twice. Multiply the circled numbers together to find the HCF.

To find the LCM write down a list of multiples of each number. Find the **smallest** number which is **common** to both lists.

Worked example

grade **C**

(a) Express 108 as a product of its prime factors.

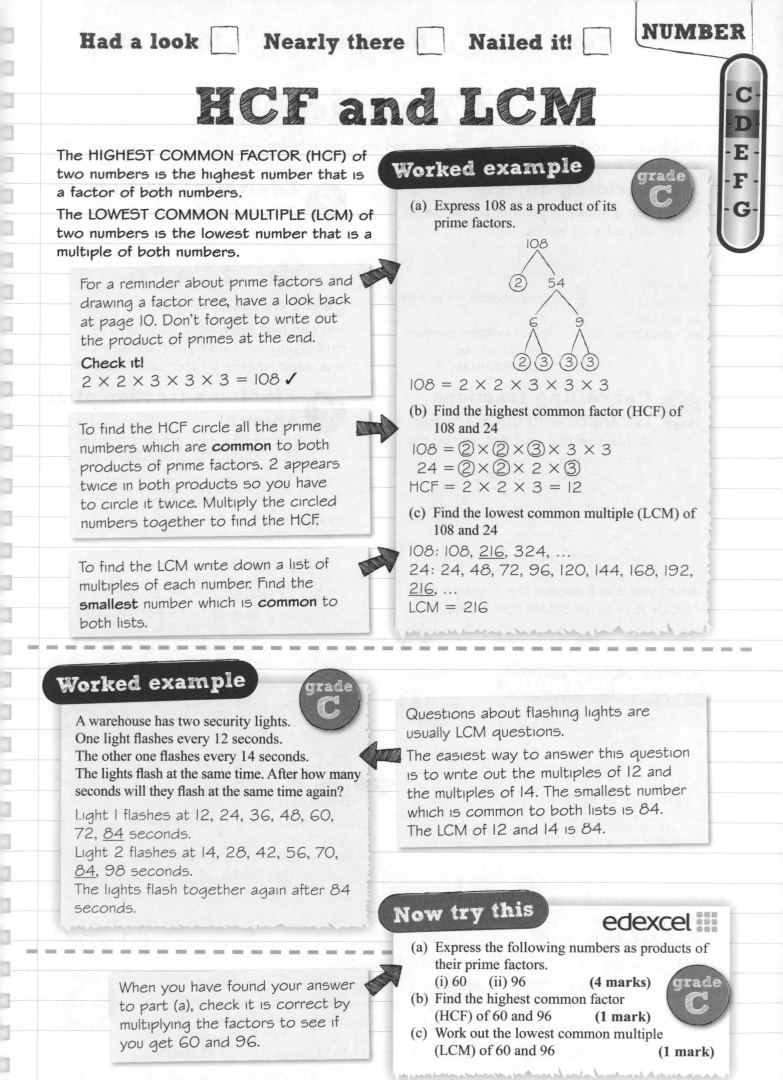

$108 = 2 \times 2 \times 3 \times 3 \times 3$

(b) Find the highest common factor (HCF) of 108 and 24

$108 = ②\times②\times③\times 3 \times 3$
$\ 24 = ②\times②\times 2 \times ③$
$HCF = 2 \times 2 \times 3 = 12$

(c) Find the lowest common multiple (LCM) of 108 and 24

108: 108, <u>216</u>, 324, ...
24: 24, 48, 72, 96, 120, 144, 168, 192, <u>216</u>, ...
LCM = 216

Worked example

grade **C**

A warehouse has two security lights.
One light flashes every 12 seconds.
The other one flashes every 14 seconds.
The lights flash at the same time. After how many seconds will they flash at the same time again?

Light 1 flashes at 12, 24, 36, 48, 60, 72, <u>84</u> seconds.
Light 2 flashes at 14, 28, 42, 56, 70, <u>84</u>, 98 seconds.
The lights flash together again after 84 seconds.

Questions about flashing lights are usually LCM questions.

The easiest way to answer this question is to write out the multiples of 12 and the multiples of 14. The smallest number which is common to both lists is 84. The LCM of 12 and 14 is 84.

Now try this

edexcel

When you have found your answer to part (a), check it is correct by multiplying the factors to see if you get 60 and 96.

(a) Express the following numbers as products of their prime factors.
   (i) 60   (ii) 96   **(4 marks)**

grade **C**

(b) Find the highest common factor (HCF) of 60 and 96   **(1 mark)**
(c) Work out the lowest common multiple (LCM) of 60 and 96   **(1 mark)**

Fractions

This page introduces fractions. Operations on simple fractions are covered on page 13, and mixed numbers are covered on page 14.

1 Dividing objects

You can use fractions to divide an object into parts.

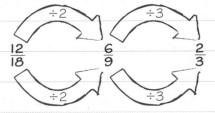

The top number is called the NUMERATOR.

$\frac{2}{3}$ of this rectangle is shaded.

The bottom number is called the DENOMINATOR.

2 Equivalent fractions

Different fractions can describe the same amount.

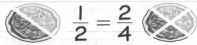

$\frac{1}{2} = \frac{2}{4}$

$\frac{1}{2}$ and $\frac{2}{4}$ are called equivalent fractions. You can find equivalent fractions by multiplying or dividing the numerator and denominator by the same number.

3 Cancelling fractions

To CANCEL or REDUCE a fraction you divide the top and bottom by the same number.

$\div 2$ $\div 3$

$\frac{12}{18}$ $\frac{6}{9}$ $\frac{2}{3}$

$\div 2$ $\div 3$

When you can't cancel the fraction any further it is in its SIMPLEST FORM.

4 Finding a fraction of an amount

Divide the amount by the denominator

Multiply by the numerator

To work out $\frac{3}{10}$ of 200 kg:

200 kg ÷ 10 = 20 kg

20 kg × 3 = 60 kg

To see how to convert between fractions and decimals see page 18.

Worked example grade G

a) Write $\frac{20}{80}$ as a fraction in its simplest form.

$\frac{20}{80} = \frac{2}{8} = \frac{1}{4}$

(b) Work out $\frac{2}{5}$ of £240

240 ÷ 5 = 48

48 × 2 = 96

$\frac{2}{5}$ of £240 is £96

When you cancel a fraction in more than one step, write down all of your steps.

$\div 10$ $\div 2$ $\div 20$

$\frac{20}{80}$ $\frac{2}{8}$ $\frac{1}{4}$ $\frac{20}{80}$ $\frac{1}{4}$

$\div 10$ $\div 2$ $\div 20$

Check it!

(a) A fraction with a numerator of 1 is in its simplest form. ✓

(b) $\frac{2}{5}$ is less than 1 so the answer should be less than £240. ✓

Now try this grade E

A school has 1200 students. 575 of these pupils are girls.

$\frac{2}{5}$ of the girls like sport.

$\frac{3}{5}$ of the boys like sport.

First work out how many of the 575 girls like sport.

Find how many of the students are boys before you work out how many of the boys like sport.

edexcel

Work out the total number of students in the school who like sport. **(3 marks)**

Simple fractions

1 Adding or subtracting

Write both fractions as equivalent fractions with the same denominator

↓

Add or subtract the numerators

↓

Do not change the denominator

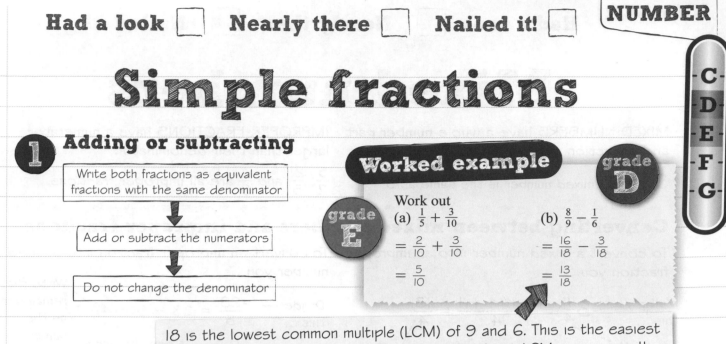

Worked example   grade **E**   grade **D**

Work out

(a) $\frac{1}{5} + \frac{3}{10}$

$= \frac{2}{10} + \frac{3}{10}$

$= \frac{5}{10}$

(b) $\frac{8}{9} - \frac{1}{6}$

$= \frac{16}{18} - \frac{3}{18}$

$= \frac{13}{18}$

18 is the lowest common multiple (LCM) of 9 and 6. This is the easiest common denominator to use. For a reminder about LCMs see page 11.

2 Multiplying

Write any whole numbers on their own as fractions with denominator 1

↓

Multiply the numerators and multiply the denominators

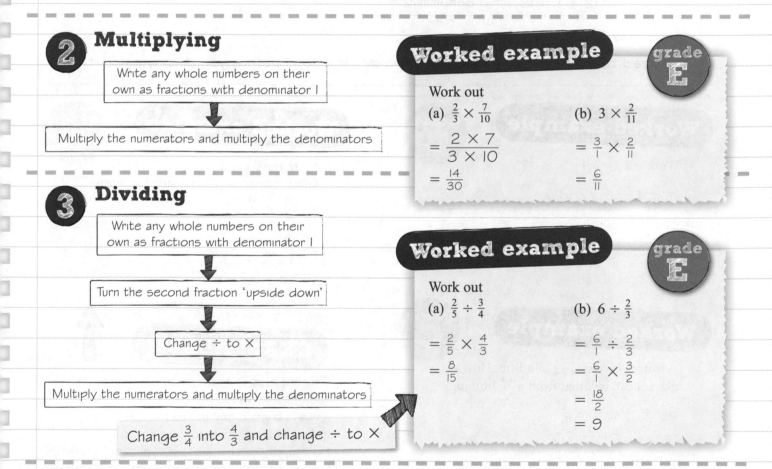

Worked example   grade **E**

Work out

(a) $\frac{2}{3} \times \frac{7}{10}$

$= \frac{2 \times 7}{3 \times 10}$

$= \frac{14}{30}$

(b) $3 \times \frac{2}{11}$

$= \frac{3}{1} \times \frac{2}{11}$

$= \frac{6}{11}$

3 Dividing

Write any whole numbers on their own as fractions with denominator 1

↓

Turn the second fraction 'upside down'

↓

Change ÷ to ×

↓

Multiply the numerators and multiply the denominators

Change $\frac{3}{4}$ into $\frac{4}{3}$ and change ÷ to ×

Worked example   grade **E**

Work out

(a) $\frac{2}{5} \div \frac{3}{4}$

$= \frac{2}{5} \times \frac{4}{3}$

$= \frac{8}{15}$

(b) $6 \div \frac{2}{3}$

$= \frac{6}{1} \div \frac{2}{3}$

$= \frac{6}{1} \times \frac{3}{2}$

$= \frac{18}{2}$

$= 9$

Watch out!

1. You do not have to cancel your final answer unless the question asks you to 'give your answer in its simplest form'.

2. You can compare and order fractions by using equivalent fractions with the same denominator.

See page 12 for a reminder about equivalent fractions and simplest form.

Now try this   edexcel ▦

(a) Work out (i) $\frac{1}{3} + \frac{3}{5}$ (ii) $\frac{11}{12} - \frac{3}{4}$

(2 marks)   grade **D**

(b) Work out (i) $\frac{2}{3} \times \frac{3}{4}$ (ii) $\frac{2}{3} \div \frac{5}{6}$

(2 marks)

Give your answers as fractions in their simplest form.

Mixed numbers

C D E F G

MIXED NUMBERS have a whole number part and a fraction part.

$3\frac{1}{4}$ This mixed number is the same as $3 + \frac{1}{4}$.

IMPROPER FRACTIONS have a numerator larger than their denominator.

$\frac{5}{2}$, $\frac{21}{5}$ and $\frac{4}{3}$ are all improper fractions.

Converting between mixed numbers and improper fractions

To convert a mixed number into an improper fraction you...

Multiply this ...

... by this...
$3 \times 4 = 12$

$3\frac{1}{4} = \frac{3 \times 4 + 1}{4} = \frac{13}{4}$

...add it to this.
$12 + 1 = 13$

Keep the same denominator

To convert an improper fraction into a mixed number you...

Divide this ...

... by this.

$\frac{23}{5} = 23 \div 5 = 4\frac{3}{5}$

Keep the same denominator

Write the remainder as the numerator

Golden rule

You need to write mixed numbers as improper fractions before you do any calculations.

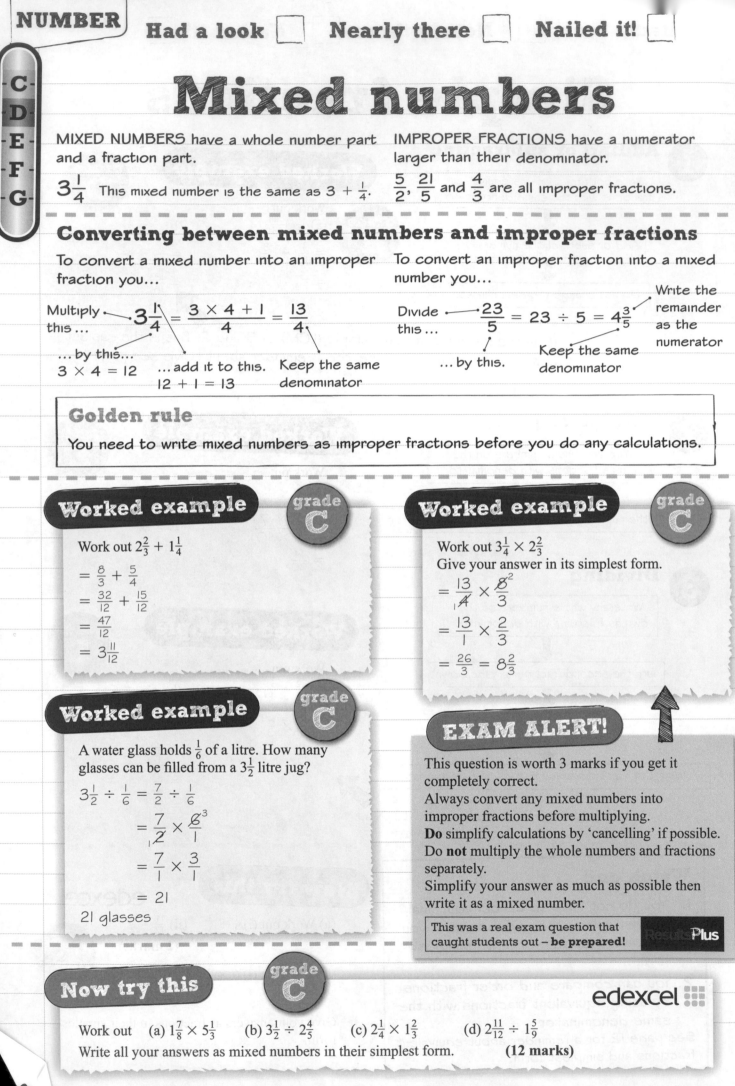

Worked example

grade C

Work out $2\frac{2}{3} + 1\frac{1}{4}$

$= \frac{8}{3} + \frac{5}{4}$

$= \frac{32}{12} + \frac{15}{12}$

$= \frac{47}{12}$

$= 3\frac{11}{12}$

Worked example

grade C

Work out $3\frac{1}{4} \times 2\frac{2}{3}$
Give your answer in its simplest form.

$= \frac{13}{\cancel{4}_1} \times \frac{\cancel{8}^2}{3}$

$= \frac{13}{1} \times \frac{2}{3}$

$= \frac{26}{3} = 8\frac{2}{3}$

Worked example

grade C

A water glass holds $\frac{1}{6}$ of a litre. How many glasses can be filled from a $3\frac{1}{2}$ litre jug?

$3\frac{1}{2} \div \frac{1}{6} = \frac{7}{2} \div \frac{1}{6}$

$= \frac{7}{\cancel{2}_1} \times \frac{\cancel{6}^3}{1}$

$= \frac{7}{1} \times \frac{3}{1}$

$= 21$

21 glasses

EXAM ALERT!

This question is worth 3 marks if you get it completely correct.
Always convert any mixed numbers into improper fractions before multiplying.
Do simplify calculations by 'cancelling' if possible.
Do **not** multiply the whole numbers and fractions separately.
Simplify your answer as much as possible then write it as a mixed number.

This was a real exam question that caught students out – **be prepared!**   ResultsPlus

Now try this

grade C

Work out   (a) $1\frac{7}{8} \times 5\frac{1}{3}$ (b) $3\frac{1}{2} \div 2\frac{4}{5}$ (c) $2\frac{1}{4} \times 1\frac{2}{3}$ (d) $2\frac{11}{12} \div 1\frac{5}{9}$

Write all your answers as mixed numbers in their simplest form.   **(12 marks)**

edexcel

Number and calculator skills

C
D
E
F
G

You should make sure you know how to use all of these functions on your calculator.

x^2 Square a number.

x^3 Cube a number.

▢ Enter a fraction. Use the down arrow to enter the bottom of the fraction.

Ans Use your previous answer in a calculation.

$(-)$ Enter a negative number.

$\sqrt{\square}$ Find the square root of a number.

$\sqrt[3]{\square}$ Find the cube root of a number. You might need to press the shift key first.

S⇔D Change the answer from a fraction or square root to a decimal.

Order of operations

You need to use the correct ORDER OF OPERATIONS when doing a calculation.

Brackets

Indices

Division

Multiplication

Addition

Subtraction

$(10 - 7) + 4 \times 3^2$
$= 3 + 4 \times 3^2$
$= 3 + 4 \times 9$
$= 3 + 36$
$= 39$

Reciprocals

To find the RECIPROCAL of a number you write it as a fraction then turn the fraction upside down.

$7 = \frac{7}{1} \rightarrow \frac{1}{7}$ The reciprocal of 7 is $\frac{1}{7}$

$\frac{3}{4} \rightarrow \frac{4}{3}$ The reciprocal of $\frac{3}{4}$ is $\frac{4}{3}$

You can use the x^{-1} key on your calculator to find reciprocals.

Worked example

grade **F**

Work out $20 - 12 \div 4$

BIDMAS
$20 - 12 \div 4$
$= 20 - 3$
$= 17$

EXAM ALERT!

This is an easy question if you remember BIDMAS, but only one-third of students got the 1 mark available for it.
You have to do the **Division** first and then the **Subtraction**. Write out a new line of working for each step.

This was a real exam question that caught students out – **be prepared!** Results Plus

Worked example

grade **E**

Use your calculator to work out the value of $(2.1 - 0.3)^2 \times 1.09$
Write down all the figures on your calculator display.

BIDMAS
$(2.1 - 0.3)^2 \times 1.09$
$= 1.8^2 \times 1.09$
$= 3.24 \times 1.09$
$= 3.5316$

You need to show your working in this question. Use these keys on your calculator to work out 1.8^2.

[1] [.] [8] [x^2] [=]

Brackets first: $2.1 - 0.3 = 1.8$
Indices next: $1.8^2 = 3.24$
Multiplication last: $3.24 \times 1.09 = 3.5316$

Now try this

edexcel

grade **E**

1. Work out
(a) $2 \times 3 + 4$
(b) $10 - 2 \times 5$
(c) $16 \div (2 \times 4)$
(d) $(12 + 9) \div 3$
(4 marks)

2. Use a calculator to work out

grade **D**

(a) $\dfrac{5.2^2 + 7.9}{8.5 - 2.364}$
(b) $\sqrt{\dfrac{4.6 - 1.58^3}{0.31}}$

Write down all the figures on your calculator display.
(6 marks)

C
D
E
F
G

Percentages

'PER CENT' means 'OUT OF 100'. You can write a percentage as a fraction over 100.

$$20\% = \frac{20}{100} = \frac{1}{5} \qquad 50\% = \frac{50}{100} = \frac{1}{2} \qquad 75\% = \frac{75}{100} = \frac{3}{4} \qquad 100\% = \frac{100}{100} = 1$$

To find a percentage of an amount:

| Divide the percentage by 100 |

⬇

| Multiply by the amount |

For example 12% of 80 cm is 9.6 cm

$12 \div 100 = 0.12$
$0.12 \times 80 = 9.6$

To write one quantity as a percentage of another:

| Divide the first quantity by the second quantity |

⬇

| Multiply your answer by 100 |

For example 3 out of 12 yoghurts in a pack are strawberry.

$3 \div 12 = 0.25$
$0.25 \times 100 = 25$

So 25% of the yoghurts are strawberry.

Non-calculator methods

You might have to work out percentages in your NON-CALCULATOR paper.

You can use multiples of 1% and 10% to calculate percentages.

To work out 10% of an amount, divide by 10.
To work out 1% of an amount, divide by 100.

When you know 10% you can work out 20%, 30%, and so on.

To work out 12% of £600:

10% of £600 is £60   600 ÷ 10 = 60

1% of £600 is £6   600 ÷ 100 = 6

So, 12% of £600 is
£60 + £6 + £6 = £72

Worked example

grade D

In a year group of 96 students,
60 own a bicycle.
Express 60 as a percentage of 96

$60 \div 96 = 0.625$
$0.625 \times 100 = 62.5$
62.5% of the students own a bicycle.

Non-calculator: $\frac{60}{96} = \frac{5}{8} = 0.625$
$= 62.5\%$

Check it!
62.5% of $96 = \frac{62.5}{100} \times 96 = 60$ ✓

Now try this

edexcel

There are 800 students at Prestfield School.
144 of these students were absent from school on Wednesday.

(a) Work out how many students were **not** absent on Wednesday. **(1 mark)**

Trudy says that more than 25% of the 800 students were absent on Wednesday.

(b) Is Trudy correct? Explain your answer. **(2 marks)**

45% of these 800 students are girls.

(c) Work out 45% of 800 **(2 marks)**

grade D

There are 176 students in Year 10.

(d) Write 176 out of 800 as a percentage. **(2 marks)**

As parts (b), (c) and (d) are worth 2 marks, make sure you show **at least** one line of working for each part.

C
D
E
F
G

Percentage change

There are two methods that can be used to increase or decrease an amount by a percentage.

Method 1

Work out 25% of £280:

$$\frac{25}{100} \times £280 = £70$$

Subtract the decrease:

£280 − £70 = £210

£280

25% OFF

Method 2

400 g

PLUS 30% EXTRA

Use a MULTIPLIER.

100% + 30% = 130%

$$\frac{130}{100} = 1.3$$

The multiplier for a 30% increase is 1.3

400 g × 1.3 = 520 g

Worked example

grade **E**

Kaz buys a car. The normal price of the car is £7200
Kaz gets a 10% <u>discount</u>.

(a) Work out 10% of £7200

$$\frac{10}{100} \times £7200 = £720$$

(b) Work out how much Kaz pays for the car.

£7200 − £720 = £6480

EXAM ALERT!

Only half of students got full marks on this question. Make sure you know that words like **discount** and **depreciation** mean that you have to **decrease** the price.

You can also use the multiplier method:

100% − 10% = 90%

$$\frac{90}{100} = 0.9$$ so the multiplier for a 10% decrease is 0.9

£7200 × 0.9 = £6480

This was a real exam question that caught students out – **be prepared!**

ResultsPlus

Worked example

grade **D**

A football club <u>increases</u> the prices of its season tickets by 4.8% each year.

In 2010 a top-price season ticket cost £550
Calculate the price of this season ticket in 2011.

$$\frac{4.8}{100} \times 550 = 26.4$$

£550 + £26.40 = £576.40

When working with money, answers must be given to 2 decimal places.

Check it!
10% of £550 is £55, so 5% is £27.50
£550 + £27.50 = £577.50, which is close to £576.40

A question may also ask you to write one quantity as a percentage of another.

For a reminder have a look at page 16.

For a reminder have a look at page 16.

Now try this

(cost of book before VAT) × 1.2 = £9.12

edexcel

grade **D**

1. The normal cost of a coat is £94

In a sale the cost of the coat is reduced by 36%.

Work out the sale price of the coat.

(3 marks)

2. Alistair sells books.

He sells each book for £9.12 including VAT at 20%.

Work out how much each book costs before VAT.

(4 marks)

grade **C**

C
D
E
F
G

Fractions, decimals and percentages

Here are three important things you need to remember for your exam.

1 To convert a fraction into a decimal you divide the numerator by the denominator.

$\frac{2}{5} = 2 \div 5 = 0.4$ $\frac{2}{3} = 0.6666... = 0.\dot{6}$

The dot tells you that the 6 repeats forever.

2 You can write any percentage as a fraction with denominator 100.

$60\% = \frac{60}{100} = \frac{6}{10} = \frac{3}{5}$

Simplify your fraction as much as possible.

3 Remember these common fraction, decimal and percentage equivalents.

Fraction	$\frac{1}{100}$	$\frac{1}{10}$	$\frac{1}{5}$	$\frac{1}{4}$	$\frac{1}{2}$	$\frac{3}{4}$
Decimal	0.01	0.1	0.2	0.25	0.5	0.75
Percentage	1%	10%	20%	25%	50%	75%

You can arrange a list of fractions, decimals and percentages in order of size by changing them to the same type.

Worked example
grade **D**

Last year, Jora spent
   30% of his salary on rent
   $\frac{2}{5}$ of his salary on entertainment
   $\frac{1}{4}$ of his salary on living expenses.
He saved the rest of his salary.
Jora spent £3600 on living expenses.
Work out how much money he saved.

$\frac{2}{5} = 40\%$ $\frac{1}{4} = 25\%$
In total Jora spent
$30\% + 40\% + 25\% = 95\%$ of his salary
This means that he saved
$100\% - 95\% = 5\%$
25% of Jora's salary = £3600
5% of Jora's salary = £3600 ÷ 5
                 = £720
Jora saved £720

EXAM ALERT!

Hardly anyone got all 5 marks for this question in the exam.

There is more than one way to answer this question.

You can convert both the fractions into percentages. This way you can see what percentage of his salary Jora saved (5%).

25% = £3600 so 5% = £3600 ÷ 5.

Alternatively, you know that $\frac{1}{4}$ = £3600 (living expenses). Use this to work out his salary (£3600 × 4 = £14 400). Then work out how much he spent on rent and entertainment. Subtract all three expenses to work out how much Jora saved.

For the second method you need to be able to find a fraction of a quantity. See page 12 if you need help.

This was a real exam question that caught students out – **be prepared!**

Result Plus

Now try this
grade **E**

1. Write these numbers in order of size. Start with the smallest number.

   0.82 $\frac{4}{5}$ 85% $\frac{2}{3}$ $\frac{7}{8}$ **(2 marks)**

   1. Convert the fractions and percentages to decimals.
   2. Write the decimals in order of size.
   3. Write each number in its original form.

2. Michelle, James and Fiona bought their mother a present.
grade **D**

   Michelle paid 40% of the total, James paid $\frac{1}{5}$ of the total and Fiona paid the rest.

   If Fiona paid £70, how much did the present cost?   **(6 marks)**

edexcel

Ratio

C
D
E
F
G

RATIOS are used to compare quantities. You can find EQUIVALENT RATIOS by multiplying or dividing by the same number.

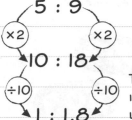

$5 : 9$

$\times 2 \qquad \times 2$

$10 : 18$

$\div 10 \qquad \div 10$

$1 : 1.8$

This equivalent ratio is in the form $1 : n$. This is useful for calculations.

Simplest form

To write a ratio in its simplest form, find an equivalent ratio with the smallest possible whole number values.

Simplest form
5 : 1   10 : 9
2 : 3 : 4

NOT simplest form
1 : 1.5 : 2   10 : 2
1 : 0.9

Worked example   grade D

There are some oranges and apples in a box.
The total number of oranges and apples is 54
The ratio of the number of oranges to the number of apples is 1 : 5
Work out the number of apples in the box.

$1 + 5 = 6$
$54 \div 6 = 9$
$5 \times 9 = 45$ apples

EXAM ALERT!

Almost 60% of students got no marks for this question.
To divide a quantity in a given ratio:
1. Work out the total number of parts in the ratio.
2. Divide the quantity by this amount.
3. Multiply your answer by each part of the ratio. Apples come after oranges in the sentence so use the second number in the ratio.

This was a real exam question that caught students out – **be prepared!**   Results **Plus**

Worked example   grade C

Jimmy is preparing the ingredients for a pizza.
He uses cheese, topping and dough in the ratio 2 : 3 : 5
He uses 70 grams of dough.
How many grams of topping will Jimmy use?

1 part: $70\,g \div 5 = 14\,g$
3 parts: $3 \times 14\,g = 42\,g$ of topping

cheese : topping : dough
  2   :   3   :   5
  ?   :   ?   :   70 g
Use 5 parts = 70 g to work out the weight of 1 part.
Multiply by 3 to find the weight of topping used.

Now try this

First work out what each person received and then find the difference.

edexcel

1. Tom is making plum crumble.

For the crumble topping, he uses sugar, butter and flour in the ratio 2 : 3 : 5

Tom uses 250 g of plums for every 400 g of crumble topping.

Tom has 0.5 kg of plums. He uses all of them.

Work out how much sugar, butter and flour he will need for the crumble topping.   **(4 marks)**

grade C

2. Alex and Ben were given a total of £360
They shared the money in the ratio 5 : 7
Ben received more money than Alex.
How much more?   **(2 marks)**

grade C

First convert 0.5 kg into grams. Then work out how much crumble topping Tom needs to make in total.
Check:
weight of sugar + butter + flour = weight of crumble topping.

Problem-solving practice

About half of the questions on your exam will need problem-solving skills.

These skills are sometimes called AO2 and AO3.

Practise these questions without a calculator. You might have to answer similar questions on your non-calculator paper.

For these questions you might need to:

- choose what method to use
- use the maths you've learnt in a new context
- plan your answer when solving harder problems
- show your working clearly and give reasons for your answers.

AO2

AO3

 Karl is carrying out a science experiment.

He records the temperature in a freezer with the door open at different times.

Here are his results.

Time	2 pm	3 pm	4 pm	5 pm
Temperature	−18 °C	−2 °C	9 °C	12 °C

Karl says that the temperature at 3 pm was halfway between the temperatures at 2 pm and 5 pm.

Is Karl correct? You must give a reason for your answer. **(2 marks)**

Negative numbers p. 8
Adding and subtracting p. 3

grade **E**

Work out the difference between −18 and −2. Then work out the difference between −2 and 12. Are they the same? You need to show **all** your working. Also remember to say whether Karl is correct or not.

TOP TIP

You can sketch a number line to help with questions about negative numbers.

 *Liam is planning a trip for a group of 20 children and 5 adults.

They can go to either the theatre or the zoo.

If they go to the theatre, they will go by train.

If they go to the zoo, they will go by coach.

Liam has information about the costs.

Theatre ticket prices	Return train fares
Stalls: £22	Adults: £11.50
Circle: £15	Child: £5.75

Zoo admission	Coach hire
Adult: £18	20 Seats: £190
Child: £12	30 Seats: £240
	40 Seats: £300

What is the lowest possible total cost of the trip? You must show all your working.

(5 marks)

Number and calculator skills p. 15

grade **E**

Work out the total cost of each trip. Remember to choose the cheapest ticket price for the theatre tickets, and write down all your working. You have to say which trip is the cheapest to complete your answer.

TOP TIP

If a question has a * next to it, then there are marks available for QUALITY OF WRITTEN COMMUNICATION. There wouldn't be an 'answer' line for this question. You must be accurate and set out all of your working neatly.

Problem-solving practice

3 Here is part of Simone's gas bill.

GAS BILL

New reading 43 214 units
Old reading 43 093 units

Price per unit 45p

Work out how much Simone has to pay for the units of gas she used.　**(4 marks)**

Number and calculator skills p. 15　grade E

There is more than one step to this question so make sure you show **all** your working clearly. You need to start by working out how many units of gas Simone used.

TOP TIP

When you are working with money you should decide whether you want to give your final answer in pounds OR in pence. DO NOT write an answer which uses both units, such as £1.99p.

---

4 A bag contains green, red and blue counters.

40% of the counters are green.

There are three times as many red counters as blue counters.

There are 9 blue counters in the bag.

How many counters are there in the bag in total?　**(4 marks)**

Percentages p. 16
Percentage change p. 17　grade D

There is more than one way to approach this question. One way is to start by working out how many red and blue counters there are in total. This represents 60% of all the counters in the bag.

TOP TIP

You can sometimes solve percentage problems by working out what 1% represents and then multiplying by 100.

---

5 Suresh wants to buy a new pair of trainers.

There are three shops that sell the trainers he wants.

Sportcentre Trainers	Footwear First Trainers	Action Sport Trainers
£10 plus 10 payments of £3.50	$\frac{1}{4}$ off usual price of £80	£40 plus VAT at 20%

Which shop is selling the trainers the cheapest?　**(5 marks)**

Fractions, decimals and percentages p. 18　grade D

There are lots of steps in this question so make sure you keep track of your working. You need to calculate the price at each shop, then write down which shop is cheapest.

TOP TIP

Divide the answer space into three columns. Then it is clear which shop each bit of working is for.

C
D
E
F
G

Collecting like terms

Expressions, equations and formulae

In algebra you use letters to represent unknown numbers.

$$4x + 3y - z$$

This is an EXPRESSION. It does not have an = sign. The parts which are separated by + or − signs are called TERMS.

$$3n - 1 = 17$$

This is an EQUATION. This equation only has one letter in it. You can solve an equation to find the value of the letter.

$$A = \tfrac{1}{2}bh$$

This is a FORMULA. You can use it to calculate one value if you know the other values. You can't solve a formula.

Simplifying expressions

You can simplify expressions which contain + and − by COLLECTING LIKE TERMS.
Like terms contain the same letters.

$h + h + h = \boxed{3h}$ → This means '3 lots of h' or $3 \times h$.

$5x - 2x = \boxed{3x}$ → '5 lots of x minus 2 lots of x equals 3 lots of x'.

$2p + 3q - 5p + q = 2p - 5p + 3q + q$
$\qquad\qquad\qquad\quad = -3p + 4q$

$2p$ and $5p$ are like terms. $2p - 5p = -3p$

Look at the opposite page for more on simplifying expressions.

Golden rules

1. Each term includes the sign (+ or −) in front of it. ✓
2. x means '1 lot of x'. You don't need to write $1x$ ✓
3. Like terms contain exactly the same combinations of letters with the same powers. ✓

Like terms
$xy \qquad -3xy$
$+10xy \qquad -xy$

NOT like terms
$3a \qquad +a^2$
$-2ab \qquad -5a^2bc$

Worked example

(a) Simplify $n + n + n + n + n$
$5n$

(b) Simplify $2p + 7p - p$
$8p$

(c) Simplify $3a + 6b - 2a + b$
$3a + 6b - 2a + b = 3a - 2a + 6b + b$
$\qquad\qquad\qquad\qquad\quad = a + 7b$

grade F

grade E

(b) Remember p means '1 lot of p'. In total there are $2 + 7 - 1 = 8$ lots of p.
(c) Combine the a terms together and combine the b terms together. Remember that each term includes the sign in front of it.

Now try this

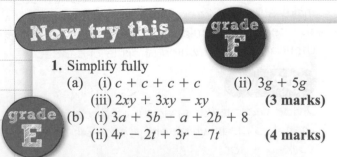

grade F

1. Simplify fully
   (a) (i) $c + c + c + c$ (ii) $3g + 5g$
       (iii) $2xy + 3xy - xy$ **(3 marks)**
   (b) (i) $3a + 5b - a + 2b + 8$
       (ii) $4r - 2t + 3r - 7t$ **(4 marks)**

grade E

2. Jamie says, '$3x - 5 = 8x$ is an equation.'
   Stuart says, '$3x - 5 = 8x$ is an expression.'
   Who is correct?
   Explain why.   **(2 marks)**

edexcel

grade E

Simplifying expressions

C
D
E
F
G

You'll need to be able to SIMPLIFY expressions which contain × and ÷ in your exam. Use these rules to help you.

① Multiplying expressions

1. Multiply any number parts first.
2. Then multiply the letters. Remember to use ☐^{2} for letters which are multiplied twice or ☐^{3} for letters which are multiplied three times.

$$10a \times 3a = 30a^2$$

$10 \times 3 = 30$ ⟋   ⟍ $a \times a = a^2$

$$3s \times 6t = 18st$$

$3 \times 6 = 18$ ⟋   ⟍ $s \times t = st$

② Dividing expressions

1. Write the division as a fraction.
2. Cancel any number parts.
3. If the same letter appears on the top and bottom, you can cancel that as well.

$$8y \div 4 = \frac{^2\cancel{8y}}{\cancel{4}_1} = 2y$$ $8 \div 4 = 2$

$$\frac{^4\cancel{36ab}}{\cancel{9b}_1} = 4a$$ $36 \div 9 = 4$

b appears on the top and the bottom, so cancel

Multiplying with algebra

You can multiply letters in algebra by writing them next to each other.
$ab = a \times b$

You can use indices to describe a letter multiplied by itself.
$y \times y = y^2$ You say 'y squared'.

You can use indices to describe the same letter multiplied together three times.
$n \times n \times n = n^3$ You say 'n cubed'.

For a reminder about squares and cubes have a look at page 9.

Worked example   grade D

(a) Simplify $a \times a \times a$

a^3

(b) Simplify $7x \times 2y$

$14xy$

(c) Simplify $10pq \div 2p$

$\frac{^5\cancel{10pq}}{\cancel{2p}_1} = 5q$

(b) Multiply the numbers first and then the letters.
$7 \times 2 = 14$ $x \times y = xy$
(c) Write the division as a fraction. You can cancel the number parts by dividing top and bottom by 2.
$10 \div 2 = 5$ so write 5 on top of your fraction.
p appears on the top and the bottom so you can cancel it. You are left with $5q$.

Now try this   grade D

Simplify
(a) $p \times p \times p \times p$ (b) $2r \times 5p$ (c) $2x \times x \times 3$
(d) $12f \div 6$ (e) $20de \div 4e$ (f) $5gh \div 5gh$

(6 marks)

Remember to look at the numbers in front of the letters first.
Then look at each letter separately.

edexcel ▦

C
D
E
F
G

Indices

Index notation

These numbers and expressions are written in index notation.

- This part is called the INDEX.
- The plural of index is indices.

12^3

- This part is called the BASE.

n^5 This means $n \times n \times n \times n \times n$.

Index laws

You can use these three index laws to simplify powers and algebraic expressions.

1 To multiply powers of the same base, add the indices.

$$a^m \times a^n = a^{m+n}$$
$$4^3 \times 4^7 = 4^{3+7} = 4^{10}$$
$$x^4 \times x^3 = x^{4+3} = x^7$$

2 To divide powers of the same base, subtract the indices.

$$a^m \div a^n = \frac{a^m}{a^n} = a^{m-n}$$
$$12^8 \div 12^3 = 12^{8-3} = 12^5$$
$$\frac{m^8}{m^2} = m^{8-2} = m^6$$

3 To raise a power of a base to a further power, multiply the indices.

$$(a^m)^n = a^{mn}$$
$$(7^3)^5 = 7^{3 \times 5} = 7^{15}$$
$$(j^2)^4 = j^{2 \times 4} = j^8$$

One at a time

When you are multiplying or dividing expressions with powers:

1. Multiply or divide any number parts first.
2. Use the index laws to work out the new power.

$7x \times 5x^6 = 35x^7$ — $x \times x^6 = x^{1+6} = x^7$

$$7 \times 5 = 35$$

$$12 \div 3 = 4$$

$\frac{12a^5}{3a^2} = 4a^3$

$$a^5 \div a^2 = a^{5-2} = a^3$$

(a) Add the indices. $a^m \times a^n = a^{m+n}$
(b) Subtract the indices. $a^m \div a^n = a^{m-n}$
(c) Multiply the indices. $(a^m)^n = a^{mn}$

Worked example

grade **C**

(a) Simplify $p^2 \times p^7$
$p^2 \times p^7 = p^{2+7} = p^9$

(b) Simplify $m^8 \div m^3$
$m^8 \div m^3 = m^{8-3} = m^5$

(c) Simplify $(a^6)^3$
$(a^6)^3 = a^{6 \times 3} = a^{18}$

Watch out!

You can only use the index laws when the bases are the same.

If there's no index then the number has a power of 1.

$$6^3 \times 6 = 6^{3+1} = 6^4$$
$$x^8 \div x = x^{8-1} = x^7$$

Now try this

grade **C**

edexcel

Simplify
(a) $p^5 \div p^3$
(b) $q \times q^2$
(c) $t^4 \times t^6$
(d) $\frac{m^9}{m^3}$
(e) $(x^4)^3$
(f) $(y^7)^2$
(g) $\frac{w^{12}}{w^4}$

(7 marks)

Use all the index laws to help you simplify.
It is a good idea to show the rule before writing the final answer. So $x^4 \times x^{11} = x^{4+11} = x^{15}$

C
D
E
F
G

Expanding brackets

Expanding brackets is sometimes called MULTIPLYING OUT brackets.

You need to be extra careful if there are negative signs outside the brackets.

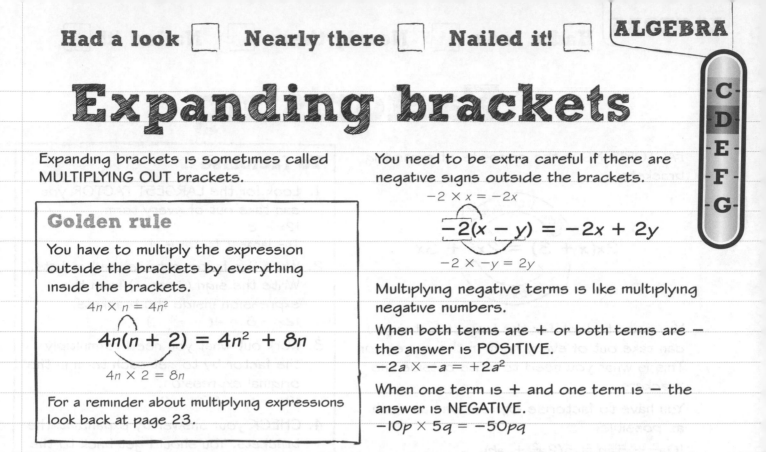

$-2 \times x = -2x$

$$-2(x - y) = -2x + 2y$$

$-2 \times -y = 2y$

Golden rule

You have to multiply the expression outside the brackets by everything inside the brackets.

$4n \times n = 4n^2$

$$4n(n + 2) = 4n^2 + 8n$$

$4n \times 2 = 8n$

For a reminder about multiplying expressions look back at page 23.

Multiplying negative terms is like multiplying negative numbers.

When both terms are + or both terms are − the answer is POSITIVE.
$-2a \times -a = +2a^2$

When one term is + and one term is − the answer is NEGATIVE.
$-10p \times 5q = -50pq$

Sometimes you have to EXPAND AND SIMPLIFY. This means 'multiply out the brackets and then collect like terms'.

$6 \times m = 6m$ $4 \times 3 = 12$

$$6(m + 2) + 4(3 - m) = 6m + 12 + 12 - 4m$$

$6 \times 2 = 12$ $4 \times -m = -4m$ $= 2m + 24$

$6m - 4m = 2m$ $12 + 12 = 24$

Remember that any negative signs belong to the term on their right.
Look back at page 22 for a reminder about collecting like terms.

Worked example

grade **C**

(a) Expand $5(2y - 3)$
$5(2y - 3) = 10y - 15$
(b) Expand and simplify $2(3x + 4) - 3(4x - 5)$
$2(3x + 4) - 3(4x - 5)$
$= 6x + 8 - 12x + 15$
$= -6x + 23$

For part (b), be careful with the second bracket.
$-3 \times 4x = -12x$
$-3 \times -5 = +15$
The question says 'expand and simplify' so remember to collect any like terms after you have multiplied out the brackets.

Now try this

grade **C**

edexcel

(a) Expand
    (i) $4(3x - 3)$ (ii) $t(3t + 4)$
    (iii) $3y(y + 4)$ **(3 marks)**
(b) Expand and simplify
    (i) $4(3m + 4) + 5(m - 5)$
    (ii) $3(2y + 4) - 2(4y - 3)$ **(4 marks)**

Be careful in part (b)(ii) as there is a minus sign outside the second bracket. Use the Worked example to help you.

Factorising

FACTORISING is the opposite of expanding brackets.

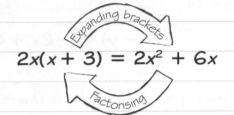

$$2x(x + 3) = 2x^2 + 6x$$

You need to look for the largest factor you can take out of every term in the expression. This is what you need to write outside the brackets.

You have to factorise expressions as much as possible.

$10a^2 + 5ab = 5(2a^2 + ab)$
This expression has only been PARTLY FACTORISED.

$10a^2 + 5ab = 5a(2a + b)$
This expression has been COMPLETELY FACTORISED.

To factorise ...

1. Look for the LARGEST FACTOR you can take out of every term.
   $12x - 8$
   The largest factor is 4.

2. Write this factor outside the brackets. Write the sign (+ or −) from the expression inside the brackets.
   $12x - 8 = 4(\quad - \quad)$

3. Work out what you need to multiply the factor by to get each term in the original expression.
   $12x - 8 = 4(3x - 2)$

4. CHECK your answer by expanding the brackets. You should get back to the original expression.

   $4 \times 3x = 12x$

   $4(3x - 2) = 12x - 8 ✓$

   $4 \times -2 = -8$

Worked example

grade D

Factorise $3x + 12$

$3x + 12 = 3(x + 4)$

EXAM ALERT!

Only 1 in 6 students got the mark for this question. Remember that 'factorise' means 'write with brackets'.

Check it!
Expanding the brackets: $3(x + 4) = 3x + 12 ✓$

This was a real exam question that caught students out – **be prepared!**   ResultsPlus

Worked example

grade C

(a) Factorise $y^2 + 3y$
$y^2 + 3y = y(y + 3)$

(b) Factorise fully $2p^2 - 4p$
$2p^2 - 4p = 2p(p - 2)$

(a) **Check it!**
$y \times y = y^2 ✓$ $y \times 3 = 3y ✓$

(b) The question says 'factorise fully'. This means there is more than one factor to take outside the brackets.
$2p^2 - 4p = 2(p^2 - 2p)$ is not fully factorised.

Now try this

grade C

edexcel

(a) Factorise
  (i) $3p - 12$ (ii) $t^2 - 5t$
  (iii) $y^2 + y$ (iv) $5 - 10v$ **(4 marks)**

(b) Factorise fully
  (i) $8a - 20$ (ii) $6b^2 + 9b$
  (iii) $14c^2 - 21c$ (iv) $10x + 25x^2$ **(4 marks)**

Always check your answer by expanding the brackets. You should get back to the original expression.

ALGEBRA

Sequences

C
D
E
F
G

Number sequences

A SEQUENCE is a pattern of numbers or shapes that follow a rule.

2, 4, 6, 8, 10, 12, ... is a sequence of even numbers.

1, 3, 5, 7, 9, 11, ... is a sequence of odd numbers.

Each number in a number sequence is called a TERM.

You can continue a sequence of numbers by finding the rule to get from one term to the next.

(a) Look for a common difference. The rule to get from one term to the next in this sequence is '− 4'.
(b) Write out all the terms up to the 10th term.
(c) Look to see if all the numbers in the sequence have anything in common. You are subtracting 4 each time so the numbers stay even.

Worked example

grade
E

Here are the first 4 terms in a number sequence.

88   84   8
−4   −4   −4

Everything in red is part of the answer.

(a) Write down the next **two** terms in this number sequence.

72, 68

(b) Write down the 10th term in this number sequence.

7th term = 64
8th term = 60
9th term = 56
10th term = 52

(c) Is 33 a member of this number sequence? Explain your answer.

No. All the numbers in the sequence are even and 33 is odd.

Pattern sequences

This is a sequence of patterns.

You can use a table to record the number of coins used in each pattern.

Pattern number 1   Pattern number 2   Pattern number 3

Pattern number	Number of coins
1	4
2	7
3	10
↓	↓
8	25

You can use the number sequence to work out how many coins will be needed to make pattern 8. You don't need to draw the pattern.

The rule to get from one term to the next is '+ 3'.
The number sequence is 4, 7, 10, 13, 16, 19, 22, 25, ... so 25 coins are needed to make pattern 8.

Now try this

edexcel

Here are some patterns made from squares.

Pattern number 1   Pattern number 2   Pattern number 3

(a) Complete the table.   **(2 marks)**

(b) Find the number of squares used for Pattern number 10   **(1 mark)**
(c) Mandy says that 130 is a number in this sequence.
    Is Mandy correct? Explain why.   **(2 marks)**

grade
E

Pattern number	1	2	3	4	5
Number of squares	5	9	13		

C
D
E
F
G

nth term of a sequence

In your exam, you may be asked to work out the nth term of a sequence. Look at this example which shows you how to do it in four steps.

1 grade **C**

Here are the first five terms of a number sequence.

1 +4 5 +4 9 +4 13 +4 17

Find, in terms of n, an expression for the nth term of the sequence.

Write in the difference between each term.

2

Here are the first five terms of a number sequence.

Zero term
−3   1 +4 5 +4 9 +4 13 +4 17

Find, in terms of n, an expression for the nth term of the sequence.

*Work backwards to find the **zero term** of the sequence. You need to subtract 4 from the first term.*

3

Here are the first five terms of a number sequence.

Zero term
−3   1 +4 5 +4 9 +4 13 +4 17

Find, in terms of n, an expression for the nth term of the sequence.

nth term = difference × n + zero term

Write down the formula for the nth term.
__Remember__ this formula for the exam.

4

Here are the first five terms of a number sequence.

Zero term
−3   1 +4 5 +4 9 +4 13 +4 17

Find, in terms of n, an expression for the nth term of the sequence.

nth term = difference × n + zero term
nth term = $4n − 3$

You can use the nth term to check whether a number is a term in the sequence.
The value of n in your nth term has to be a POSITIVE whole number.
So is 99 a term of the sequence?
Try some different values of n:
when n = 25, $4n − 3 = 4 × 25 − 3 = 97$
when n = 26, $4n − 3 = 4 × 26 − 3 = 101$
You can't use a value of n between 25 and 26 so 99 isn't a term in the sequence.
TIP!
If 99 is a term of the sequence then $4n − 3 = 99$.
You can solve this equation to show that n is not a whole number which proves that 99 is not in the sequence.

Check it!

Check your answer by substituting values of n into your nth term.
1st term: when n = 1,
$4n − 3 = 4 × 1 − 3 = 1$ ✓
2nd term: when n = 2,
$4n − 3 = 4 × 2 − 3 = 5$ ✓
You can also generate any term of the sequence.
For the 20th term, n = 20:
$4n − 3 = 4 × 20 − 3 = 77$
So the 20th term is 77.

Now try this   grade **C**

edexcel

(a) Here are the first 5 terms of a number sequence: 3   7   11   15   19
   Write down an expression, in terms of n, for the nth term of this sequence.   **(2 marks)**

(b) Adeel says that 318 is a term in the number sequence.
   Is Adeel correct? You must justify your answer.   **(2 marks)**

(c) The nth term of a different number sequence is $5 − 2n$
   Work out the first three terms of this number sequence.   **(3 marks)**

Equations 1

An EQUATION is like a pair of scales. The equals sign tells you that the scales are BALANCED. The letter represents an unknown weight. You can solve the equation to find the value of the letter.

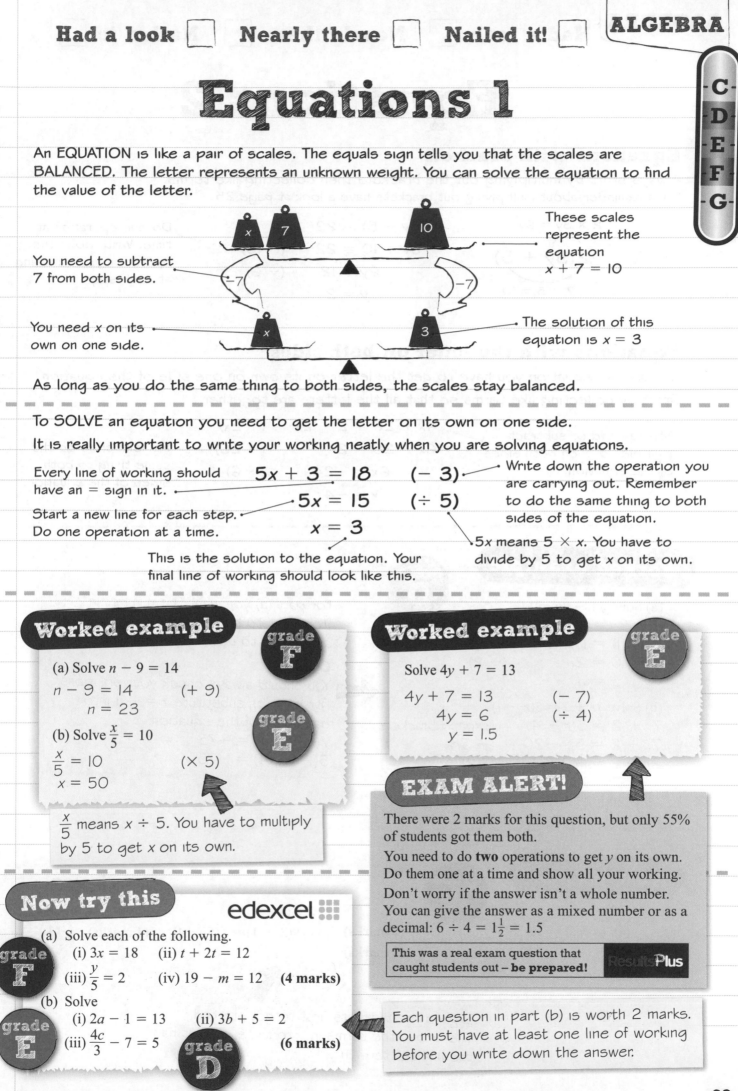

These scales represent the equation $x + 7 = 10$

You need to subtract 7 from both sides.

You need x on its own on one side.

The solution of this equation is $x = 3$

As long as you do the same thing to both sides, the scales stay balanced.

To SOLVE an equation you need to get the letter on its own on one side.

It is really important to write your working neatly when you are solving equations.

Every line of working should have an = sign in it.

Start a new line for each step. Do one operation at a time.

$$5x + 3 = 18 \qquad (- 3)$$
$$5x = 15 \qquad (\div 5)$$
$$x = 3$$

Write down the operation you are carrying out. Remember to do the same thing to both sides of the equation.

$5x$ means $5 \times x$. You have to divide by 5 to get x on its own.

This is the solution to the equation. Your final line of working should look like this.

Worked example

grade **F**

grade **E**

(a) Solve $n - 9 = 14$

$n - 9 = 14 \qquad (+ 9)$
$\quad n = 23$

(b) Solve $\frac{x}{5} = 10$

$\frac{x}{5} = 10 \qquad (\times 5)$
$x = 50$

$\frac{x}{5}$ means $x \div 5$. You have to multiply by 5 to get x on its own.

Worked example

grade **E**

Solve $4y + 7 = 13$

$4y + 7 = 13 \qquad (- 7)$
$\quad 4y = 6 \qquad (\div 4)$
$\quad\quad y = 1.5$

EXAM ALERT!

There were 2 marks for this question, but only 55% of students got them both.

You need to do **two** operations to get y on its own. Do them one at a time and show all your working.

Don't worry if the answer isn't a whole number. You can give the answer as a mixed number or as a decimal: $6 \div 4 = 1\frac{1}{2} = 1.5$

This was a real exam question that caught students out – **be prepared!**   ResultsPlus

Now try this

edexcel ▦

grade **F**

(a) Solve each of the following.
 (i) $3x = 18$ (ii) $t + 2t = 12$
 (iii) $\frac{y}{5} = 2$ (iv) $19 - m = 12$ (**4 marks**)

grade **E**

(b) Solve
 (i) $2a - 1 = 13$ (ii) $3b + 5 = 2$
 (iii) $\frac{4c}{3} - 7 = 5$

grade **D**

(**6 marks**)

Each question in part (b) is worth 2 marks. You must have at least one line of working before you write down the answer.

C
D
E
F
G

Equations 2

Equations with brackets

Always start by multiplying out the brackets then collecting like terms.
For a reminder about multiplying out brackets have a look at page 25.

$2 \times 3y = 6y$

$2(3y + 5)$

$2 \times 5 = 10$

$2(3y + 5) = 22$

$6y + 10 = 22 \quad (- 10)$

$6y = 12 \quad (\div 6)$

$y = 2$

Do one operation at a time. Write down the operation you are using at each step.

Equations with the letter on both sides

To solve an equation you have to get the letter on its own on one side of the equation.
Start by collecting like terms so that all the letters are together.

You can add or subtract multiples of x on both sides of the equation.

$4x + 26 = 2 - 2x \quad (+ 2x)$

$6x + 26 = 2 \quad (- 26)$

$6x = -24 \quad (\div 6)$

$x = -4$

Remember to do the same thing to both sides of the equation.

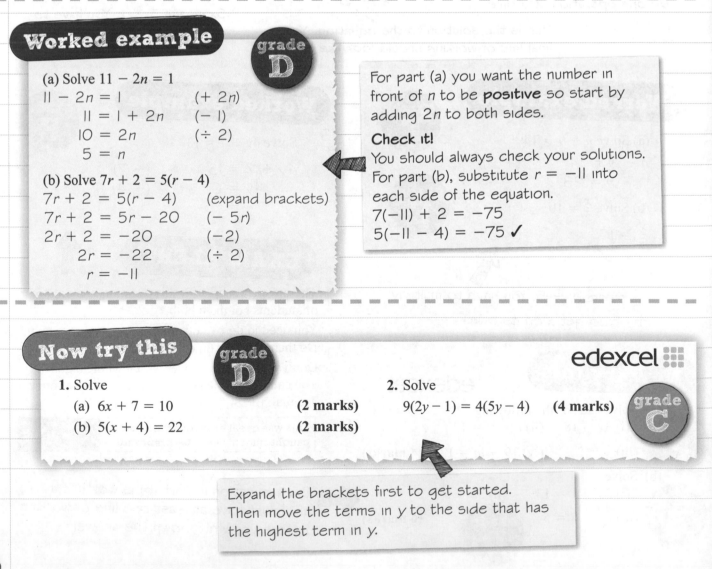

Worked example   grade D

(a) Solve $11 - 2n = 1$

$11 - 2n = 1 \quad (+ 2n)$

$11 = 1 + 2n \quad (- 1)$

$10 = 2n \quad (\div 2)$

$5 = n$

(b) Solve $7r + 2 = 5(r - 4)$

$7r + 2 = 5(r - 4) \quad \text{(expand brackets)}$

$7r + 2 = 5r - 20 \quad (- 5r)$

$2r + 2 = -20 \quad (-2)$

$2r = -22 \quad (\div 2)$

$r = -11$

For part (a) you want the number in front of n to be **positive** so start by adding 2n to both sides.

Check it!
You should always check your solutions.
For part (b), substitute $r = -11$ into each side of the equation.

$7(-11) + 2 = -75$

$5(-11 - 4) = -75 ✓$

Now try this   grade D

edexcel

1. Solve

   (a) $6x + 7 = 10$ (2 marks)

   (b) $5(x + 4) = 22$ (2 marks)

2. Solve

   $9(2y - 1) = 4(5y - 4)$ (4 marks)   grade C

Expand the brackets first to get started.
Then move the terms in y to the side that has the highest term in y.

Writing equations

C
D
E
F
G

You might need to use the information given in a question to write an equation. You might need to choose a letter to represent an unknown quantity you are trying to find.

Worked example

grade **C**

Kasam is 34 years older than his daughter.
He is also 3 times as old as she is.
Find Kasam's age.

$$x + 34 = 3x \qquad (- x)$$
$$34 = 2x \qquad (\div 2)$$
$$17 = x$$

Kasam's daughter is 17 years old.
17 + 34 = 51
Kasam is 51 years old.

You can solve this problem quickly by writing an equation. Choose a letter to represent the age of Kasam's daughter (x).

Kasam's age can be expressed as $x + 34$ (34 years older than his daughter) and as $3x$ (3 times as old as his daughter). Both of these are expressions for his age and so must be equal to each other. Solve the equation to find the age of Kasam's daughter, and then work out Kasam's age.

Worked example

grade **C**

The perimeter of this
quadrilateral is
65 cm.
Work out the
value of r.

$2r + 5$
$2r$
r
$4r - 3$

Perimeter of quadrilateral
$$= 2r + 2r + 5 + r + 4r - 3$$
$$= 9r + 2$$
$$9r + 2 = 65 \qquad (- 2)$$
$$9r = 63 \qquad (\div 9)$$
$$r = 7 \, cm$$

If a diagram uses letters to represent sides or angles then try writing an equation.
1. Write an **expression** for the perimeter of the quadrilateral in terms of r.
2. Collect like terms.
3. The perimeter is 65 cm. Use this information to turn your expression into an **equation**.
4. Solve your equation to find the value of r.

Now try this

For part (b) remember to add all four sides of the large rectangle, in terms of x, and then simplify.
In part (c) you can put your expression from part (b) equal to 20 to solve for x.

edexcel

Flowerbed   x

2

5   x

Diagram
NOT
accurately
drawn

The diagram represents a garden in the shape of a rectangle.
All measurements are given in metres.

The garden has a flowerbed in one corner.
The flowerbed is a square of side x.

grade **C**

(a) Write down an expression, in terms of x, for the shorter side of the garden.
(1 mark)

(b) Find an expression, in terms of x, for the perimeter of the garden.
Give your answer in its simplest form.
(2 marks)

The perimeter of the garden is 20 metres.
(c) Find the value of x.   **(2 marks)**

31

Had a look ☐ Nearly there ☐ Nailed it! ☐

Trial and improvement

Some equations can't be solved exactly. You need to use trial and improvement to find an approximate solution. You will be told when to use trial and improvement in your exam. Look at the worked example below which shows you how to do it in two steps.

Worked example

1 The equation $x^3 - 5x = 60$ has a solution between 4 and 5
Use a trial and improvement method to find this solution.
Give your answer correct to 1 decimal place.

x	$x^3 - 5x$	Too big or too small
4.5	68.625	Too big
4.2	53.088	Too small

Draw a table to record your working. You know there is a solution between 4 and 5.

$x = 4.5$ is a good first value to try.

Use your calculator to work out $4.5^3 - 5 \times 4.5$ and compare your answer with 60.

$x = 4.5$ is too big.

Try $x = 4.2$

2 The equation $x^3 - 5x = 60$ has a solution between 4 and 5
Use a trial and improvement method to find this solution.
Give your answer correct to 1 decimal place.

x	$x^3 - 5x$	Too big or too small
4.5	68.625	Too big
4.2	53.088	Too small
4.3	58.007	Too small
4.4	63.184	Too big
4.35	60.56...	Too big

$x = 4.3$ (to 1 d.p.)

Keep trying different values.

Make sure you write down the result of every trial.

You know the answer is between 4.3 and 4.4. But you don't know which value is closer.

Try 4.35. This will tell you whether the answer is closer to 4.3 or 4.4

4.35 is too big so the answer is between 4.3 and 4.35

Write down the answer correct to 1 decimal place.

Now try this

1. The equation $x^3 + 2x = 26$ has a solution between 2 and 3

   Use a trial and improvement method to find this solution.

   Give your answer correct to 1 decimal place.

   You must show **all** your working.

   (4 marks)

2. The equation $x^3 + 4x^2 = 100$ has a solution between 3 and 4

   Use a trial and improvement method to find this solution.

   Give your answer correct to 1 decimal place.

   You must show **all** your working.

   (4 marks)

edexcel

Inequalities

You can use these symbols to describe INEQUALITIES. The fat end of the symbol always points towards the bigger number.

$>$ means 'is greater than'
$x > 2$

$\geq$ means 'is greater than or equal to'
$y \geq -3$

$<$ means 'is less than'
$a < 1$

$\leq$ means 'is less than or equal to'
$b \leq -9$

Worked example grade C

(a) Write $>$ or $<$ between each pair of numbers to make a true statement.
  (i) $7 \boxed{<} 10$ (ii) $-2 \boxed{>} -6$
(b) Write down all the values of x that are whole numbers and satisfy the inequality
  $-1 \leq x < 3$

  $x = -1, 0, 1, 2$

(a) (ii) -2 is to the right of -6 on a number line, so $-2 > -6$.
(b) A number **satisfies** an inequality if it makes a true statement. -1 is included but 3 isn't.

Inequalities on number lines

You can represent inequalities on a number line.

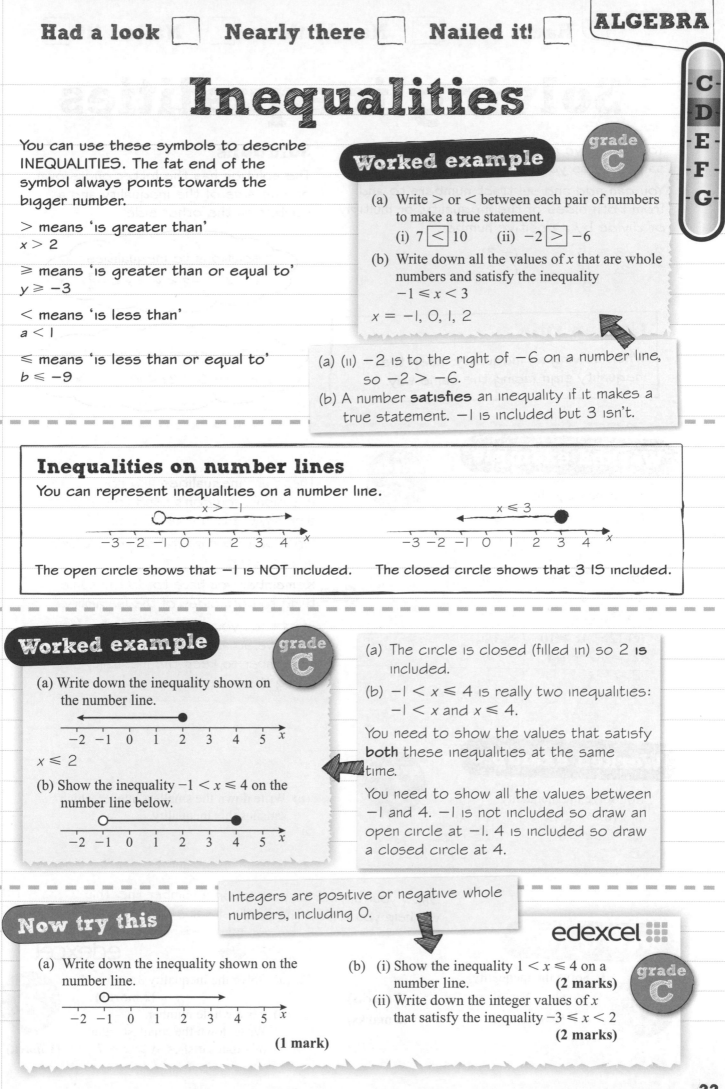

The open circle shows that -1 is NOT included.   The closed circle shows that 3 IS included.

Worked example grade C

(a) Write down the inequality shown on the number line.

$x \leq 2$

(b) Show the inequality $-1 < x \leq 4$ on the number line below.

(a) The circle is closed (filled in) so 2 is included.

(b) $-1 < x \leq 4$ is really two inequalities: $-1 < x$ and $x \leq 4$.

You need to show the values that satisfy **both** these inequalities at the same time.

You need to show all the values between -1 and 4. -1 is not included so draw an open circle at -1. 4 is included so draw a closed circle at 4.

Integers are positive or negative whole numbers, including 0.

Now try this

(a) Write down the inequality shown on the number line.

(1 mark)

edexcel

(b) (i) Show the inequality $1 < x \leq 4$ on a number line. **(2 marks)**
    (ii) Write down the integer values of x that satisfy the inequality $-3 \leq x < 2$
    (2 marks)

grade C

C
D
E
F
G

Solving inequalities

You can solve an inequality in exactly the same way as you solve an equation.

You can add and subtract numbers to and from both sides of the inequality, or multiply or divide by a positive number.

$$2x - 3 \leq 15 \qquad (+ 3)$$
$$\quad\ 2x \leq 18 \qquad (\div 2)$$
$$\qquad x \leq 9$$

Golden rule

When solving an inequality, keep the inequality sign facing the same way.

Solutions

The solution has the letter on its own on one side of the inequality and a number on the other side.

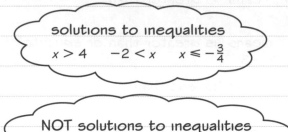

solutions to inequalities

$$x > 4 \qquad -2 < x \qquad x \leq -\tfrac{3}{4}$$

NOT solutions to inequalities

$$x \geq 20 + 3 \qquad 2x < 10 \qquad x = 4$$

Worked example

grade C

Solve the inequality

(a) $2x + 1 < 10 \qquad (- 1)$
$\quad\ 2x < 9 \qquad\quad (\div 2)$
$\qquad x < \tfrac{9}{2} \qquad\quad$ (or $x < 4\tfrac{1}{2}$)

(b) $5x < 2x - 6 \qquad (- 2x)$
$\quad\ 3x < -6 \qquad\quad (\div 3)$
$\qquad x < -2$

(c) $12 - 4x \geq 10 \qquad (+ 4x)$
$\quad 12 \geq 10 + 4x \qquad (- 10)$
$\qquad 2 \geq 4x \qquad\quad (\div 4)$
$\qquad \tfrac{2}{4} \geq x \qquad\quad$ (or $\tfrac{1}{2} \geq x$)

These are **inequalities** and not equations. You don't need to use an = sign anywhere in your answer.

Write down the operation you are using at each step.

Remember you have to do the same thing to both sides of the inequality.

In part (c), you don't want 4x to be negative so add 4x to both sides. Remember to keep the inequality sign facing the same way.

Worked example

grade C

(a) Solve the inequality
$\qquad 10 - x < 2x + 1 \qquad (+ x)$
$\quad 10 < 3x + 1 \qquad\quad (- 1)$
$\qquad 9 < 3x \qquad\qquad (\div 3)$
$\qquad 3 < x$

(b) Write down the smallest integer which satisfies this inequality.

4

Show your answer to part (a) on a number line. This will help you to find the smallest value of x in part (b).

Now try this

grade C

1. Solve the inequality
   (a) $6x - 3 \leq 9$ **(2 marks)**
   (b) $5y + 17 > 2$ **(2 marks)**

2. (a) Solve the inequality $3x \geq x - 7$
   (2 marks)
   (b) x is a whole number.
         Write down the smallest value
         of x that satisfies $3x \geq x - 7$ **(1 mark)**

grade C

edexcel

Substitution

If you know the values of the letters in an algebraic expression, you can SUBSTITUTE them into the expression. This lets you work out the value of the expression.

$x = 7$ and $y = 2$ have been substituted into this expression.

$x \quad + \quad 5y$ — 5y means $5 \times y$. When $y = 2$, $5y = 5 \times 2 = 10$

$x = 7$ $\quad$ $y = 2$

$7 + 5 \times 2 = 7 + 10$
$\qquad \qquad = 17$ — Multiply before you Add.

You should substitute all the values before doing any calculations.

When $x = 7$ and $y = 2$ the value of $x + 5y$ is 17.

Using BIDMAS

Remember to use the correct order of operations when you are doing a calculation. BIDMAS gives the order in which the operations should be carried out.

Brackets

Indices

Division

Multiplication

Addition

Subtraction

$2 \times 5^2 - (14 + 8)$
$= 2 \times 5^2 - 22$
$= 2 \times 25 - 22$
$= 50 - 22$
$= 28$

For more on BIDMAS have a look at page 15.

Worked example

grade **D**

Be extra careful when substituting a negative number. You can use brackets around the number to make sure you don't make a mistake.

(a) Work out the value of $5x + 1$ when $x = -3$
$5 \times (-3) + 1 = -15 + 1$
$\qquad \qquad \qquad = -14$

(b) Work out the value of $5p^3$ when $p = 2$
$5 \times 2^3 = 5 \times 8$
$\qquad \quad = 40$

Remember **BIDMAS**. Indices comes before Multiplication.

Substitute all the values before starting your calculation.

(c) Work out the value of $3m + 4n$ when $m = 5$ and $n = -2$
$3 \times 5 + 4 \times (-2) = 15 + (-8)$
$\qquad \qquad \qquad \quad = 15 - 8$
$\qquad \qquad \qquad \quad = 7$

(d) Work out the value of $2x(x - 1)$ when $x = 11$
$2 \times 11 \times (11 - 1) = 2 \times 11 \times 10$
$\qquad \qquad \qquad \qquad = 22 \times 10$
$\qquad \qquad \qquad \qquad = 220$

You can multiply in any order.
$22 \times 10 = 2 \times 110$

Now try this

grade **D**

1. (a) Work out the value of $3x - 4y$ when
$x = 3$ and $y = 2$ **(2 marks)**

(b) $p = 3t + 4(q - t)$
Find the value of p when $q = 8$ and $t = 5$
(2 marks)

(c) Work out the value of $\dfrac{p(q - 3)}{4}$ when
$p = 2$ and $q = -7$ **(2 marks)**

2. Tayub said, 'When $x = 3$, then the value of $4x^2$ is 144'
Bryani said, 'When $x = 3$, then the value of $4x^2$ is 36'
Who was right?
Explain why. **(2 marks)**

grade **D**

edexcel

C
D
E
F
G

Formulae

A FORMULA is a mathematical rule.

Formulae is the plural of formula.

The formula for the area of this triangle is:

Area = $\frac{1}{2}$ × base × height

You can write this formula using algebra as:

$A = \frac{1}{2}bh$

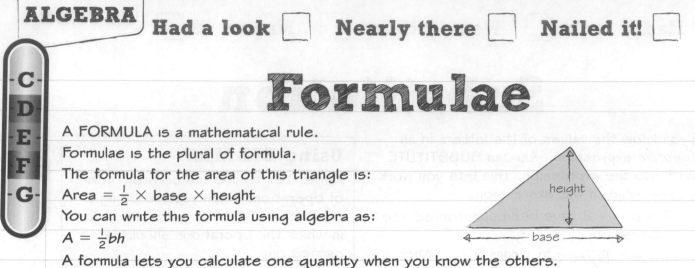

A formula lets you calculate one quantity when you know the others.

You need to SUBSTITUTE the values you know into the formula.

For more on substitution have a look at page 35.

Worked example

grade F

Andrea uses this formula to work out her gas bill.

Cost = fixed charge + cost per unit × units used

Last month her fixed charge was £40
She used 62 units. The cost per unit was 50p.
How much was Andrea's gas bill last month?

$50p = £0.50$

Cost = £40 + £0.50 × 62
 = £40 + £31
 = £71

When calculating using formulae make sure all your units are the same — either pounds or pence when working in money. It makes sense to work in pounds here so write 50p in pounds.

Substitute all your values before calculating the cost.

Remember to use **BIDMAS** for the correct order of operations. You have to **M**ultiply before you **A**dd.

For a reminder about BIDMAS have a look at page 35.

Substitute the values for *u* and *t* into the formula. Using **BIDMAS**:

1. Do the **I**ndex (power) first.
   $3^2 = 9$
2. Do the **M**ultiplications next.
   $20 \times 3 = 60$
   $5 \times 9 = 45$
3. Do the **S**ubtraction last.
   $60 - 45 = 15$

Don't try to do more than one operation on each line of working.

Worked example

grade D

This formula is used in physics to calculate distance.

$D = ut - 5t^2$
$u = 20$
$t = 3$

Work out the value of *D*.

$D = 20 \times 3 - 5 \times 3^2$
 $= 20 \times 3 - 5 \times 9$
 $= 60 - 45$
 $= 15$

Now try this

 edexcel

grade D

This rule is used to work out the cost of hiring a carpet cleaner.

$C = 4D + 6$

where *C* is the cost in £ and *D* is the number of days the carpet cleaner is hired.

Peter hires the carpet cleaner for 4 days.

(a) Work out how much this will cost him. **(1 mark)**

Another company uses this rule to work out the cost of hiring a drill.

$C = 3(H + 2)$

where *C* is the cost in £ and *H* is the number of hours the drill is hired.

(b) Work out *C* when $H = 3$ **(1 mark)**

C
D
E
F
G

Writing formulae

You can write a rule given in words as a word formula or as a formula using algebra.

This label gives instructions for working out the cooking time of a chicken.

FREE-RANGE CHICKEN		
WEIGHT (KG) 1.8	PRICE PER KG £3.95	COOKING INSTRUCTIONS Cook at 170°C for 25 minutes per kg plus half an hour

You can write the cooking instruction as a word formula.

Cooking time in minutes = 25 × weight in kg + 30

You need to give units when you are describing the quantities in a formula.
If the cooking time is in hours then this formula would give you a very crispy chicken!

You can also write this formula using algebra.

$T = 25w + 30$, where T is the cooking time in minutes and w is the weight in kg.
When you write a formula using algebra you need to explain what each letter means.

Worked example   grade D

Tom the plumber charges £35 for each hour he works at a job, plus £50
His total fee can be worked out using this rule:

> Multiply number of hours by 35
> Add 50 to your answer

Tom works for h hours and is paid P pounds. Write down a formula for P in terms of h.

$P = 35h + 50$

Remember $35h$ means $35 × h$.
Do not write any units (e.g. £) in your formula.

Worked example   grade D

The cost of hiring a car is £100 plus £50 for each hire day.

Rita hires a car for d days.

Write a formula for the total hire cost £C in terms of d.

Cost = £100 + £50 × number of days
$C = 100 + 50d$

Sometimes it helps to write a formula in words first, then write it using algebra.

Worked example   grade D

The diagram shows a regular hexagon.
Write a formula for the perimeter of the hexagon P in terms of l.

$P = 6l$

←l→

The perimeter is the distance all the way around a shape.
All the sides of a regular shape are the same length. There are 6 sides so:
perimeter = 6 × length of one side
$P = 6l$

You can find out more about perimeter on page 65.

Now try this

Write down your formula showing what you have added before simplifying.
For a reminder about simplifying have a look at page 23.

edexcel

grade C

In the diagram, all measurements are in cm.

$x + 6$
$2x - 3$
$3x + 1$

Diagram **NOT** accurately drawn

(a) Write down a formula for the perimeter, P, of the triangle in terms of x.
Give your formula in its simplest form.
(2 marks)

(b) Work out the value of P when $x = 5$ cm.
(1 mark)

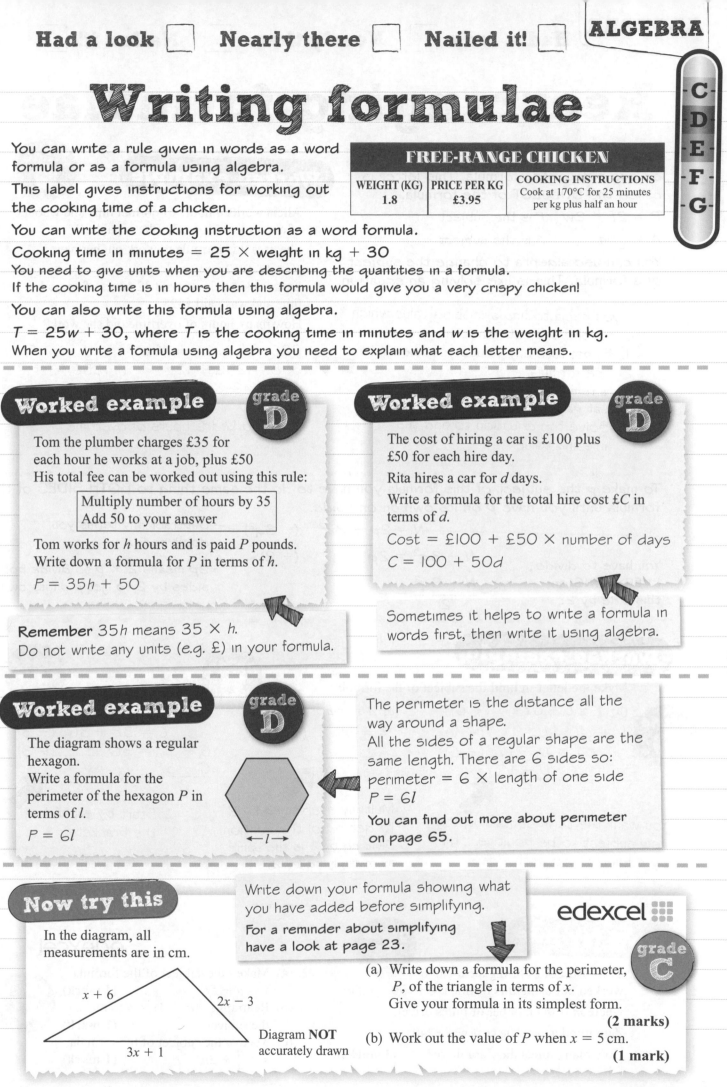

Rearranging formulae

Most formulae have one letter on its own on one side of the formula. This letter is called the SUBJECT of the formula.

$P = 2l + 2w$ P is the subject

$A = \frac{1}{2}bh$ A is the subject

You can use algebra to change the subject of a formula. This is like solving an equation.

If you need to find a missing value which is not the subject in a formula:
1. Substitute any values you know into the formula.
For a reminder about substitution have a look at page 35.
2. Solve the equation to find the missing value.

Worked example

grade **D**

Alicia works h hours of normal time and v hours of overtime each week. She is paid £P

She uses this formula to work out her pay:

$P = 8h + 12v$

Last week Alicia worked 32 hours of normal time and was paid £328

How many hours of overtime did she work?

$328 = 8 \times 32 + 12v$
$328 = 256 + 12v$ $(- 256)$
$72 = 12v$ $(\div 12)$
$6 = v$

Alicia worked 6 hours of overtime

To make p the subject of this formula you have to do the same thing to BOTH SIDES of the formula until you have p on its own on one side.

$$N = 2p + 3q^2 \quad (- 3q^2)$$ — Subtract any terms you don't need.

$$N - 3q^2 = 2p \quad (\div 2)$$

You have to divide EVERYTHING on this side by 2.

$$\frac{N - 3q^2}{2} = p$$

$2p$ means $2 \times p$ so divide both sides by 2 to get p on its own.

Worked example

Make the letter in **bold** the subject of the formula.

grade **D**

(a) $V = U - \mathbf{GT}$ $(+ GT)$
$V + GT = U$ $(- V)$
$GT = U - V$ $(\div G)$
$T = \frac{U - V}{G}$

You want the term containing the new subject to be positive. Start by adding GT to both sides.

grade **D**

(b) $A = \frac{1}{2}\mathbf{b}h$ $(\times 2)$
$2A = bh$ $(\div b)$
$\frac{2A}{b} = h$

Multiply everything by 2 to get rid of the fraction. Dividing by $\frac{1}{2}$ is the same as multiplying by 2.

grade **C**

(c) $C = 3(P + \mathbf{Q})$
$C = 3P + 3Q$ $(- 3P)$
$C - 3P = 3Q$ $(\div 3)$
$\frac{C - 3P}{3} = Q$

Start by expanding the brackets.

Now try this

grade **D**

1. The cost, £C, of hiring a car can be worked out using this formula: $C = 90 + 0.5m$
where m is the number of miles driven.
Zara hired a car. The cost was £240
How many miles did Zara drive?   **(3 marks)**

2. (a) Make t the subject of the formula
$v = u + 5t$ **(1 mark)**

grade **C**

(b) Rearrange $a(q - 3) = d$ to make q the subject.   **(1 mark)**
(c) Make a the subject of the formula
$s = \frac{a}{4} + 2u$ **(1 mark)**

edexcel

Coordinates

You can use coordinates to describe the positions of points on a grid.

This point is to the left of O on the x-axis. So the x-coordinate is negative.

The vertical axis is labelled y.

The first number in a coordinate pair describes the horizontal position. The second number describes the vertical position.

The horizontal axis is labelled x.

You can use negative numbers to describe points below O on the y-axis.

The point O is called the origin and has coordinates (0, 0).

$(-1, 2)$ $(3, 2)$ $(-4, -1)$ $(0, -2)$ $(2, -4)$

Worked example

grade **F**

(a) Write down the coordinates
   (i) of the point A
   (2, 3)
   (ii) of the point B
   (2, −2)

(b) On the grid, plot the point D so that ABCD is a rectangle.

Everything in red is part of the answer.

A rectangle has four right angles. Plot point D level with point A and vertically above point C.

Midpoints

A LINE SEGMENT is a short section of a straight line. The midpoint of a line segment is exactly halfway along the line. You can find the midpoint if you know the coordinates of the ends.

To find the midpoint, add the x-coordinates and divide by 2 and add the y-coordinates and divide by 2.

$$\text{Midpoint} = \left(\frac{x_1 + x_2}{2}, \frac{y_1 + y_2}{2}\right)$$

$(12, 5)$

Midpoint $(8, 3)$

$(4, 1)$

$\frac{1 + 5}{2} = \frac{6}{2} = 3$

$\frac{4 + 12}{2} = \frac{16}{2} = 8$

You need to be confident in finding midpoints if you're going for a grade D or C.

Now try this

edexcel

(a) Write down the coordinates of the point P. **(1 mark)**
(b) On the grid, mark the point (5, −2) with a cross (×).
    Label the point R. **(1 mark)**

PQRS is a parallelogram.

(c) Write down the coordinates of S. **(2 marks)**
(d) Write down the coordinates of the midpoint of the line PQ. **(2 marks)**

grade **F**

grade **D**

C D E F G

Straight-line graphs 1

The GRADIENT of a straight-line graph measures how steep the line is. You can work out the gradient by drawing a triangle and using this rule: Gradient = $\dfrac{\text{distance up}}{\text{distance across}}$

Worked example

grade D

This scatter graph shows the relationship between the budgets of some films and the amounts of money they made at the box office in their opening weekends.
A line of best fit has been drawn on the scatter graph.

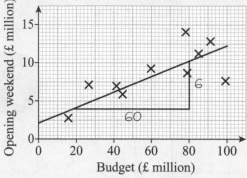

Work out the gradient of the line of best fit.

Gradient = $\dfrac{\text{distance up}}{\text{distance across}} = \dfrac{6}{60} = 0.1$

Everything in red is part of the answer.

A straight line on a graph has the equation $y = mx + c$. There is more on $y = mx + c$ on the opposite page.

To work out the gradient of a line you need to draw a triangle.

Write the distance across and the distance up.

Watch out for the scales on the axes:

Distance across = 80 – 20 = 60

Distance up = 10 – 4 = 6

Top triangle tips!

1. Draw one side of your triangle on a large grid line as you are less likely to make a mistake in your calculations.

2. Use a large triangle as this means your calculations are more accurate.

3. Don't just count grid squares. Use the scale to work out the distance across and the distance up.

Positive or negative?

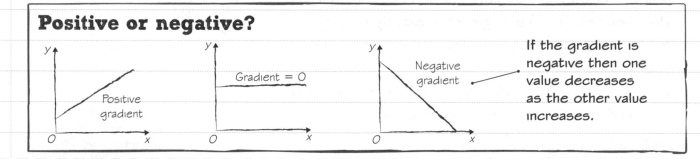

If the gradient is negative then one value decreases as the other value increases.

Now try this

grade C   edexcel

The point A has coordinates (0, 3).
The point B has coordinates (2, 7).

(a) Work out the gradient of the line that passes through A and B.
(2 marks)

C is the point with coordinates (7, 1).

(b) Work out the gradient of the line that passes through B and C.
(2 marks)

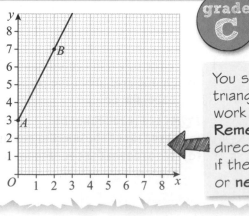

You should **draw** the triangles on the grid to work out the gradient.
Remember to look at the direction of the line to see if the gradient is **positive** or **negative**.

Straight-line graphs 2

C
D
E
F
G

A straight line on a graph has the equation

$$y = mx + c$$

where m and c are numbers.

c is the y-intercept. This is the point where the line crosses the y-axis.

m is the gradient. This measures how steep the line is.

For the straight line $y = 3x + 1$, the gradient is 3 and the y-intercept is (0, 1).

Horizontal and vertical

Horizontal lines have the equation '$y = a$' and vertical lines have the equation '$x = a$', where a is a number.

Worked example

grade **C**

On the grid draw the graph of $x + y = 4$ for values of x from -2 to 5

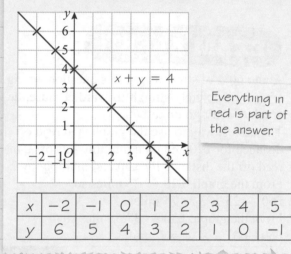

$x + y = 4$

Everything in red is part of the answer.

x	-2	-1	0	1	2	3	4	5
y	6	5	4	3	2	1	0	-1

EXAM ALERT!

Three quarters of students got no marks on this question.

To draw a straight-line graph:
1. Make a table of values. Choose simple values for x. You might be told which values of x to use. Substitute each value for x into the equation and find the value of y that makes it true.
   When $x = -2$: $-2 + y = 4 \rightarrow y = 6$
   When $x = 1$: $1 + y = 4 \rightarrow y = 3$
2. Plot the points from the table on the grid. The first point is $(-2, 6)$.
3. Join the points with a straight line using a ruler.

This was a real exam question that caught students out – **be prepared!**

ResultsPlus

Finding the equation of a line

To find the equation of a line you need to know the gradient and the y-intercept.

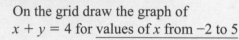

Draw a triangle to find the gradient.

Gradient $= \dfrac{20}{4} = 5$

The y-intercept is (0, 20).

Put your values for gradient, m, and y-intercept, c, into the equation of a straight line, $y = mx + c$.

The equation is $y = 5x + 20$

Now try this

edexcel

(a) On the grid, draw the graph of $y = 3x + 1$

(3 marks)

grade **C**

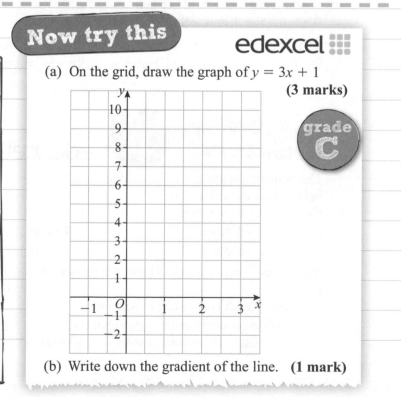

(b) Write down the gradient of the line. **(1 mark)**

C
D
E
F
G

Real-life graphs

You can use a graph to convert between different units or currencies.

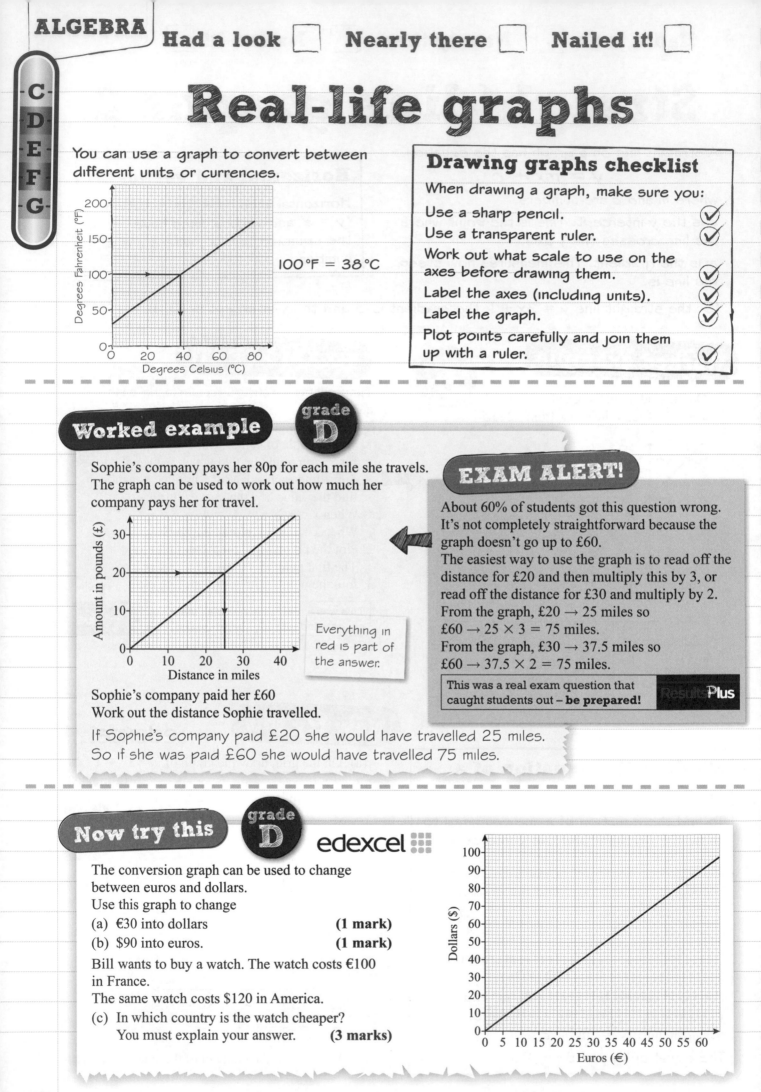

$100\,°F = 38\,°C$

Drawing graphs checklist

When drawing a graph, make sure you:

Use a sharp pencil. ✓

Use a transparent ruler. ✓

Work out what scale to use on the axes before drawing them. ✓

Label the axes (including units). ✓

Label the graph. ✓

Plot points carefully and join them up with a ruler. ✓

Worked example   grade D

Sophie's company pays her 80p for each mile she travels. The graph can be used to work out how much her company pays her for travel.

Everything in red is part of the answer.

Sophie's company paid her £60
Work out the distance Sophie travelled.

If Sophie's company paid £20 she would have travelled 25 miles.
So if she was paid £60 she would have travelled 75 miles.

EXAM ALERT!

About 60% of students got this question wrong. It's not completely straightforward because the graph doesn't go up to £60.
The easiest way to use the graph is to read off the distance for £20 and then multiply this by 3, or read off the distance for £30 and multiply by 2.
From the graph, £20 → 25 miles so
£60 → 25 × 3 = 75 miles.
From the graph, £30 → 37.5 miles so
£60 → 37.5 × 2 = 75 miles.

This was a real exam question that caught students out – **be prepared!**   Results Plus

Now try this   grade D   edexcel

The conversion graph can be used to change between euros and dollars.
Use this graph to change
(a) €30 into dollars   **(1 mark)**
(b) $90 into euros.   **(1 mark)**

Bill wants to buy a watch. The watch costs €100 in France.
The same watch costs $120 in America.
(c) In which country is the watch cheaper?
    You must explain your answer.   **(3 marks)**

Distance-time graphs

A DISTANCE-TIME graph shows how distance changes with time. This distance-time graph shows Jodi's run. The shape of the graph gives you information about the journey.

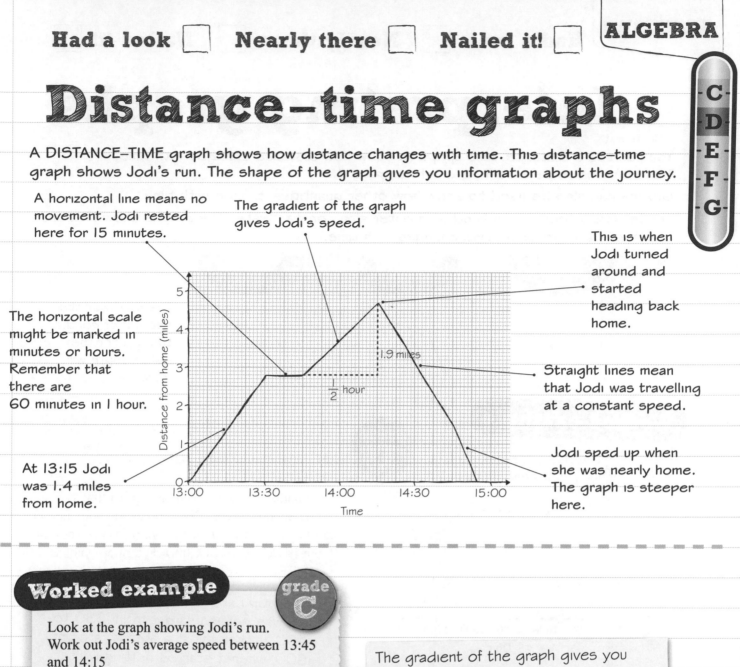

A horizontal line means no movement. Jodi rested here for 15 minutes.

The gradient of the graph gives Jodi's speed.

This is when Jodi turned around and started heading back home.

The horizontal scale might be marked in minutes or hours. Remember that there are 60 minutes in 1 hour.

Straight lines mean that Jodi was travelling at a constant speed.

At 13:15 Jodi was 1.4 miles from home.

Jodi sped up when she was nearly home. The graph is steeper here.

Worked example

grade C

Look at the graph showing Jodi's run.
Work out Jodi's average speed between 13:45 and 14:15

Jodi travelled 1.9 miles in $\frac{1}{2}$ hour.
$\frac{1}{2}$ hour = 0.5 hour
$1.9 \div 0.5 = 1.9 \times 2 = 3.8$
Jodi was travelling at 3.8 mph on this section of the run.

The gradient of the graph gives you Jodi's speed.

$\text{Gradient} = \dfrac{\text{distance up}}{\text{distance across}} = 3.8$

There is more about calculating speed on page 64.

Now try this

edexcel

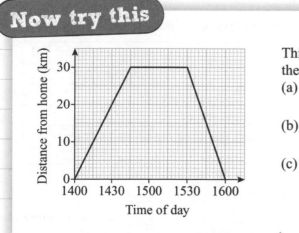

This distance-time graph shows Judy's journey to the airport and back.

(a) What is the distance from Judy's home to the airport? **(1 mark)**

(b) For how many minutes did Judy wait at the airport? **(1 mark)**

(c) Work out Judy's average speed on her journey home from the airport.
Give your answer in kilometres per hour. **(2 marks)**

grade D

grade C

Interpreting graphs

You can use a distance–time graph to show how distance changes with time. There is more on distance–time graphs on page 43.

Graphs can also be used to show how other quantities change with time.

These garden ponds are filled with water at a constant rate. The graphs show how the depth of water in each pond changes with time.

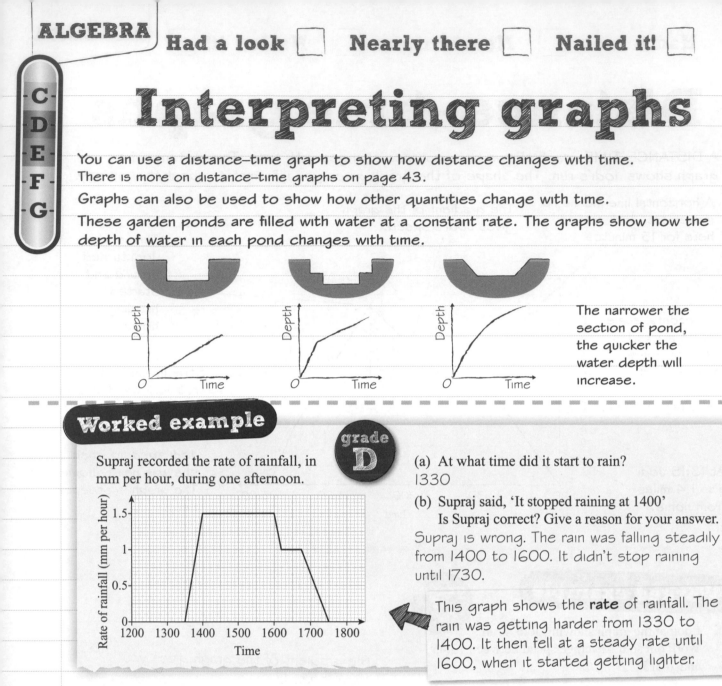

The narrower the section of pond, the quicker the water depth will increase.

Worked example

grade **D**

Supraj recorded the rate of rainfall, in mm per hour, during one afternoon.

(a) At what time did it start to rain?
1330

(b) Supraj said, 'It stopped raining at 1400'
Is Supraj correct? Give a reason for your answer.
Supraj is wrong. The rain was falling steadily from 1400 to 1600. It didn't stop raining until 1730.

This graph shows the **rate** of rainfall. The rain was getting harder from 1330 to 1400. It then fell at a steady rate until 1600, when it started getting lighter.

For a reminder about other real-life graphs, e.g. conversion graphs, have a look at page 42.

Now try this

grade **D**

Here are six temperature–time graphs.

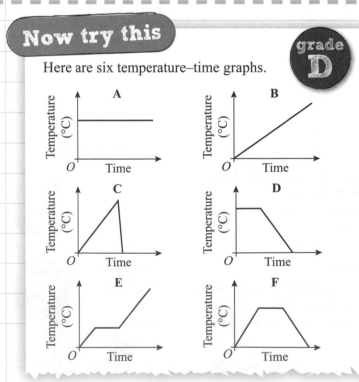

edexcel

Each sentence in the table describes the temperature for one of the graphs.

Write the letter of the correct graph next to each sentence.

The first one has been done for you.

It starts at 0 °C and keeps rising.	**B**
It stays the same for a time and then falls.	
It rises and then falls quickly.	
It is always the same.	
It rises, stays the same for a time and then falls.	
It rises, stays the same for a time and then rises again.	

(3 marks)

C
D
E
F
G

Quadratic graphs

An equation which contains an x^2 term is called a QUADRATIC equation. Quadratic equations have CURVED graphs. You can draw the graph of a quadratic equation by completing a table of values.

Worked example

grade **C**

(a) Complete the table of values for $y = 4x - x^2$

x	−1	0	1	2	3	4	5
y	−5	0	3	4	3	0	−5

(b) On the grid, draw the graph of $y = 4x - x^2$

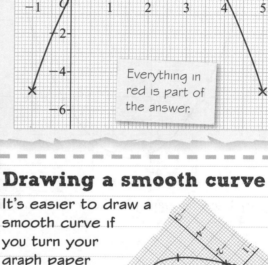

Substitute each value of x into the equation to get a corresponding value of y.

When $x = -1$: $4 \times -1 - (-1)^2 = -4 - 1 = -5$
When $x = 4$: $4 \times 4 - 4^2 = 16 - 16 = 0$

Plot your points carefully on the graph and join them with a **smooth** curve.

Check it!

All the points on your graph should lie on the curve. If one of the points doesn't fit then double-check your calculation.

Everything in red is part of the answer.

Drawing quadratic curves

Use a sharp pencil. ✓

Plot the points carefully. ✓

Draw a smooth curve that passes through every point. ✓

Label your graph. ✓

Shape of graph will be either $\smile$ or $\frown$

Drawing a smooth curve

It's easier to draw a smooth curve if you turn your graph paper so your hand is INSIDE the curve.

Now try this

(a) Complete the table of values for
$y = x^2 + 2x - 3$ **(2 marks)**

x	−4	−3	−2	−1	0	1	2
y		0	−3			0	5

(b) On the grid, draw the graph of
$y = x^2 + 2x - 3$ **(2 marks)**

Be careful when substituting. Remember that $(-4)^2 = 16$

edexcel

grade **C**

Using quadratic graphs

An equation with an x^2 in it is called a QUADRATIC equation.
You can find SOLUTIONS to a quadratic equation by drawing a graph.

Worked example

grade **C**

(a) Complete the table of values for $y = x^2 - x - 3$

x	-2	-1	0	1	2	3
y	3	-1	-3	-3	-1	3

(b) On the grid, draw the graph of $y = x^2 - x - 3$

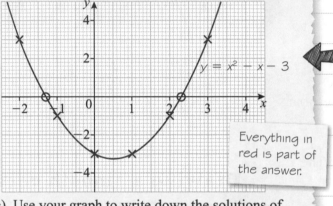

$y = x^2 - x - 3$

Everything in red is part of the answer.

In part (b) you have drawn the graph of $y = x^2 - x - 3$.

$x^2 - x - 3 = 0$ is the equation of the graph with $y = 0$ substituted in.

The solutions of the equation are the points where the graph crosses the x-axis.

For every point on the x-axis, $y = 0$.

For the solutions, read your graph to the nearest small square.

Remember that quadratic equations have **two** solutions.

(c) Use your graph to write down the solutions of
$x^2 - x - 3 = 0$

$x = -1.3$ and $x = 2.3$

Solving quadratic equations

This is a graph of $y = 2x^2 + 5x$

You can use this graph to find the values of x when $y = 2$

1. Draw a horizontal line at $y = 2$

2. Read down to the x-axis at the points where the line crosses the curve.

3. Write your answers to 1 decimal place.

The values of x are 0.4 and -2.9
These are the solutions to the equation $2x^2 + 5x = 2$

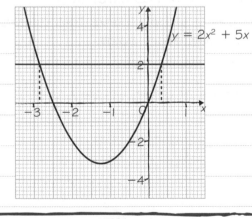

$y = 2x^2 + 5x$

Now try this

edexcel

(a) Complete the table of values for
$y = x^2 - 4x - 2$ **(2 marks)**

x	-1	0	1	2	3	4	5
y		-2	-5			-2	3

(b) Draw a grid with the x-axis from $+5$ to -1 and y-axis from $+4$ to -6
On the grid, draw the graph of
$y = x^2 - 4x - 2$ **(2 marks)**

grade **C**

(c) Use your graph to estimate the values of x when $y = -3$ **(2 marks)**

(d) Use your graph to write down the solutions of
$x^2 - 4x - 2 = 0$ **(2 marks)**

Problem-solving practice

About half of the questions on your exam will need problem-solving skills.

These skills are sometimes called AO2 and AO3.

You can use a calculator on question 3, but practise the other questions without one.
You might have to answer similar questions on your non-calculator paper.

For these questions you might need to:

* choose what method to use
* use the maths you've learnt in a new context
* plan your answer when solving harder problems
* show your working clearly and give reasons for your answers.

AO2
AO3

---

1 Here are some patterns made from dots.

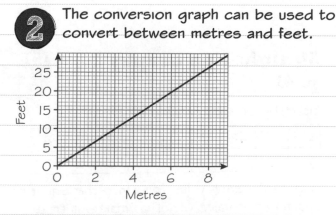

Pattern number 1   Pattern number 2   Pattern number 3

(a) How many dots are needed for Pattern number 5?   **(1 mark)**

(b) Vidya says she can make a pattern out of 21 dots.

   Is Vidya correct? You must give a reason for your answer.   **(2 marks)**

Sequences p. 27

grade E

For part (a) you should draw a table showing the number of dots in each pattern. Carry on the number sequence until you get to Pattern number 5. For part (b) you should look to see if 21 could be a number in the sequence. If it can't, make sure you give a reason why not.

TOP TIP

Check your number sequence by drawing the next pattern and counting the dots.

---

2 The conversion graph can be used to convert between metres and feet.

Jenny needs 150 feet of ribbon.

The ribbon costs £2 per metre.

Work out the cost of the ribbon Jenny needs.   **(3 marks)**

Real-life graphs p. 42

grade E

Start by converting 15 feet into metres. Draw a line across from 15 on the vertical axis to the line. Then draw a line down to the horizontal axis. Use this value to do the conversion needed.

Be careful when you are reading off the value from the horizontal axis. Here, 10 small squares represents 2 m, so 1 small square represents 0.2 m.

TOP TIP

When reading off values from a conversion graph, draw on the lines you use with a ruler.

Problem-solving practice

3 Dhruv uses this formula to work out how long to cook a chicken.

$$\text{Cooking time in minutes} = \text{chicken's weight in kg} \times 45 + 30$$

He works out that his chicken will take exactly 1 hour 33 minutes to cook.

Work out the weight of Dhruv's chicken.

(3 marks)

Rearranging formulae p. 38 — grade D

The units in the formula are minutes so write 1 hour 33 minutes as a time in minutes.
To find the weight of Dhruv's chicken, substitute any values you know into the formula, and then solve the equation to find the missing value.

TOP TIP

You can represent the chicken's weight in kg as a single letter, e.g. w, in order to simplify your equation.

4

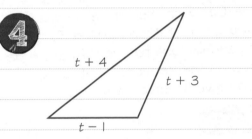

The perimeter of this triangle is 19 cm.
All lengths on the diagram are in cm.
Work out the value of t. (4 marks)

Writing equations p. 31 — grade D

Use the information in the question to write an equation. Solve your equation to work out the value of t.

TOP TIP

Remember to simplify your equation by collecting like terms before solving.

---

5 On the grid, draw the graph of $x + y = 5$ for values of x from -3 to 4

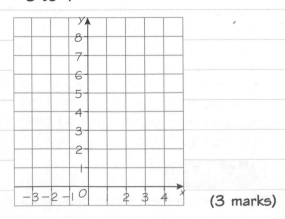

(3 marks)

Straight-line graphs 2 p. 41 — grade C

Start by drawing a table of values.

x	-3	-2	-1	0	1	2	3	4
y							2	

Substitute each value for x into the equation and find the value of y that makes it true.

TOP TIP

It's easier to start with the positive values of x. Look to see if the values of y in the table follow a pattern.

C
D
E
F
G

Measuring and drawing angles

Measuring angles

1 A protractor measures angles in degrees.

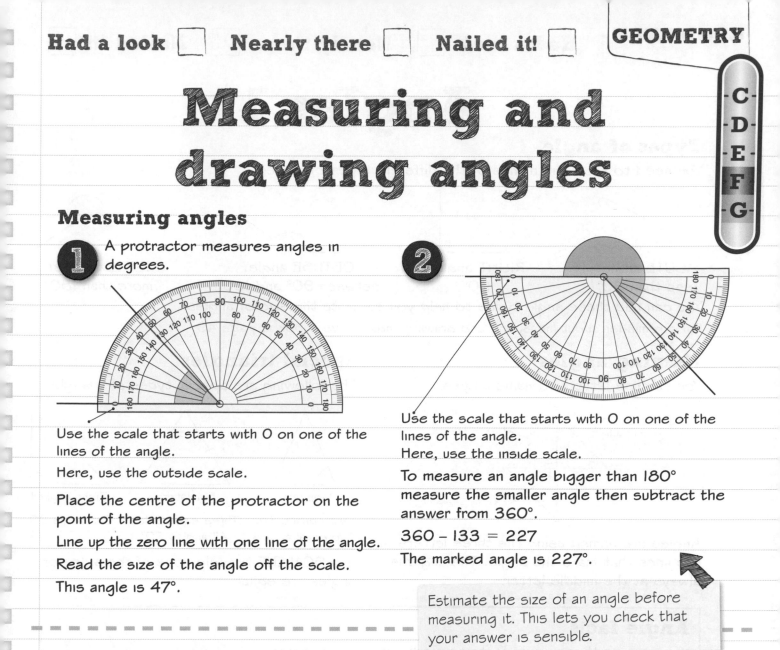

Use the scale that starts with 0 on one of the lines of the angle.

Here, use the outside scale.

Place the centre of the protractor on the point of the angle.

Line up the zero line with one line of the angle.

Read the size of the angle off the scale.

This angle is 47°.

2

Use the scale that starts with 0 on one of the lines of the angle.

Here, use the inside scale.

To measure an angle bigger than 180° measure the smaller angle then subtract the answer from 360°.

360 − 133 = 227

The marked angle is 227°.

> Estimate the size of an angle before measuring it. This lets you check that your answer is sensible.

Drawing angles

1 Draw an angle of 23°.

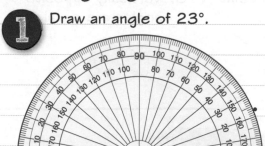

Use the scale that starts with 0 on one of the lines of the angle.

Here, use the inside scale.

Use a ruler to draw one line of the angle, AB.

Place the centre of your protractor on one end of the line. The zero line needs to lie along your line.

Find 23° on the scale. Draw a dot to mark this point.

2

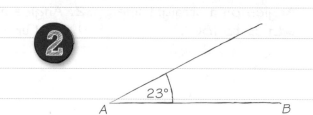

Use a ruler to join the end of the line and your dot with a straight line.

Draw in the angle curve and label your angle 23°.

Now try this

edexcel

Measure the size of each of these angles.
(a) **(1 mark)**   (b) **(1 mark)**

grade **F**

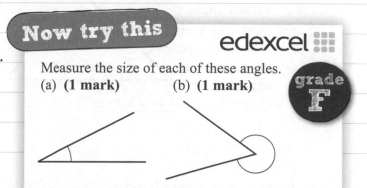

Angles 1

Types of angle

You need to know the names of the different types of angles.

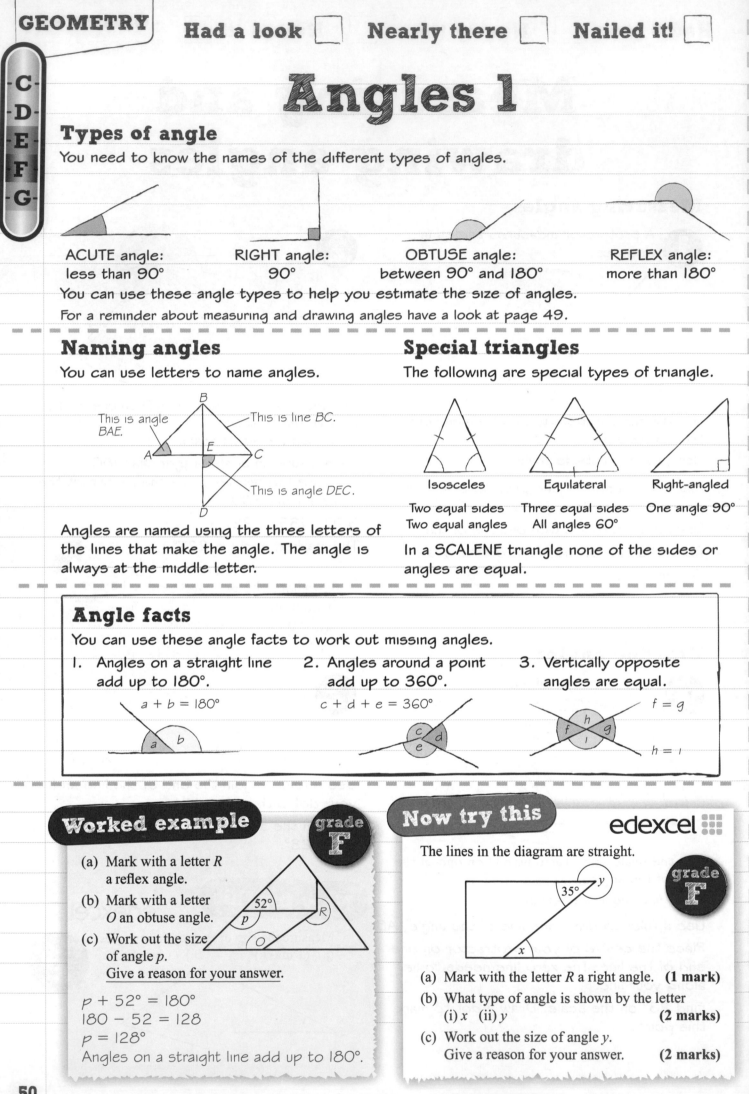

ACUTE angle:
less than 90°

RIGHT angle:
90°

OBTUSE angle:
between 90° and 180°

REFLEX angle:
more than 180°

You can use these angle types to help you estimate the size of angles.

For a reminder about measuring and drawing angles have a look at page 49.

Naming angles

You can use letters to name angles.

This is angle BAE.

This is line BC.

This is angle DEC.

Angles are named using the three letters of the lines that make the angle. The angle is always at the middle letter.

Special triangles

The following are special types of triangle.

Isosceles
Two equal sides
Two equal angles

Equilateral
Three equal sides
All angles 60°

Right-angled
One angle 90°

In a SCALENE triangle none of the sides or angles are equal.

Angle facts

You can use these angle facts to work out missing angles.

1. Angles on a straight line add up to 180°.

   $a + b = 180°$

2. Angles around a point add up to 360°.

   $c + d + e = 360°$

3. Vertically opposite angles are equal.

   $f = g$

   $h = i$

Worked example

grade F

(a) Mark with a letter *R* a reflex angle.

(b) Mark with a letter *O* an obtuse angle.

(c) Work out the size of angle *p*.
Give a reason for your answer.

$p + 52° = 180°$
$180 - 52 = 128$
$p = 128°$
Angles on a straight line add up to 180°.

Now try this

edexcel

The lines in the diagram are straight.

35°

grade F

(a) Mark with the letter *R* a right angle.   **(1 mark)**

(b) What type of angle is shown by the letter
(i) *x* (ii) *y*   **(2 marks)**

(c) Work out the size of angle *y*.
Give a reason for your answer.   **(2 marks)**

Angles 2

Triangles and quadrilaterals

These are useful facts for triangles and quadrilaterals.

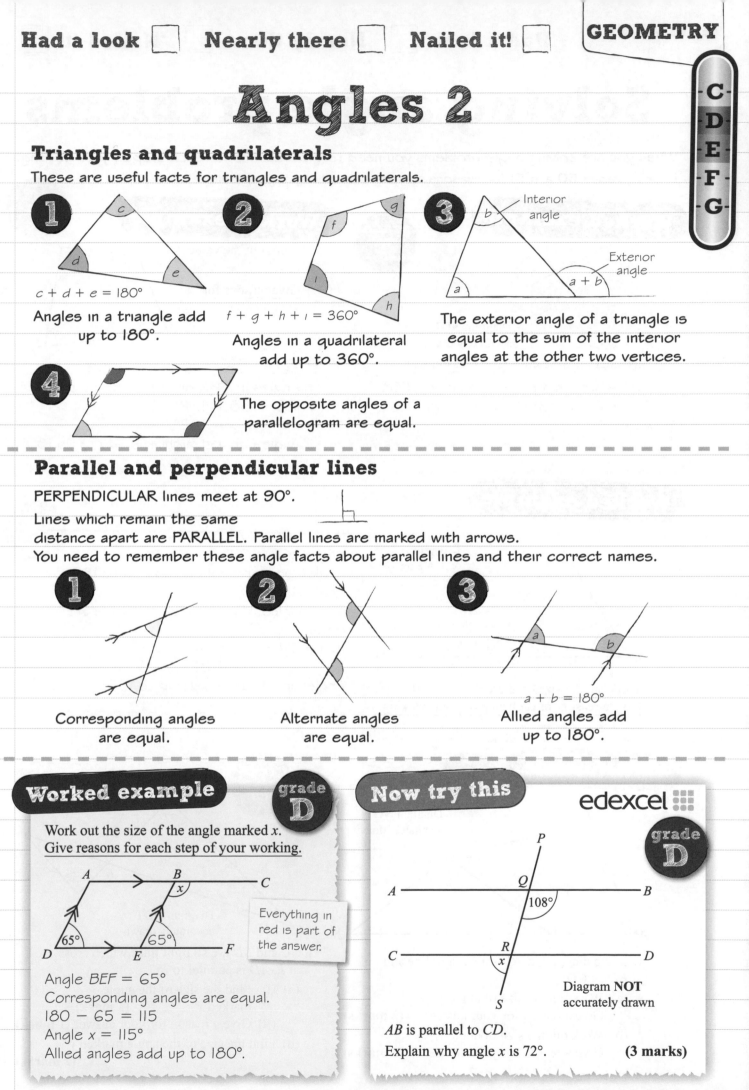

1

$c + d + e = 180°$

Angles in a triangle add up to 180°.

2

$f + g + h + i = 360°$

Angles in a quadrilateral add up to 360°.

3

Interior angle

Exterior angle

The exterior angle of a triangle is equal to the sum of the interior angles at the other two vertices.

4

The opposite angles of a parallelogram are equal.

Parallel and perpendicular lines

PERPENDICULAR lines meet at 90°.

Lines which remain the same distance apart are PARALLEL. Parallel lines are marked with arrows.

You need to remember these angle facts about parallel lines and their correct names.

1

Corresponding angles are equal.

2

Alternate angles are equal.

3

$a + b = 180°$

Allied angles add up to 180°.

Worked example

grade **D**

Work out the size of the angle marked *x*.
Give reasons for each step of your working.

> Everything in red is part of the answer.

Angle *BEF* = 65°
Corresponding angles are equal.
180 − 65 = 115
Angle *x* = 115°
Allied angles add up to 180°.

Now try this

edexcel

grade **D**

108°

x

Diagram **NOT** accurately drawn

AB is parallel to *CD*.
Explain why angle *x* is 72°.   **(3 marks)**

C
D
E
F
G

Solving angle problems

When you are solving angle problems you need to give reasons for each step of your working.
Look at pages 50 and 51 for reasons to use when solving angle problems.

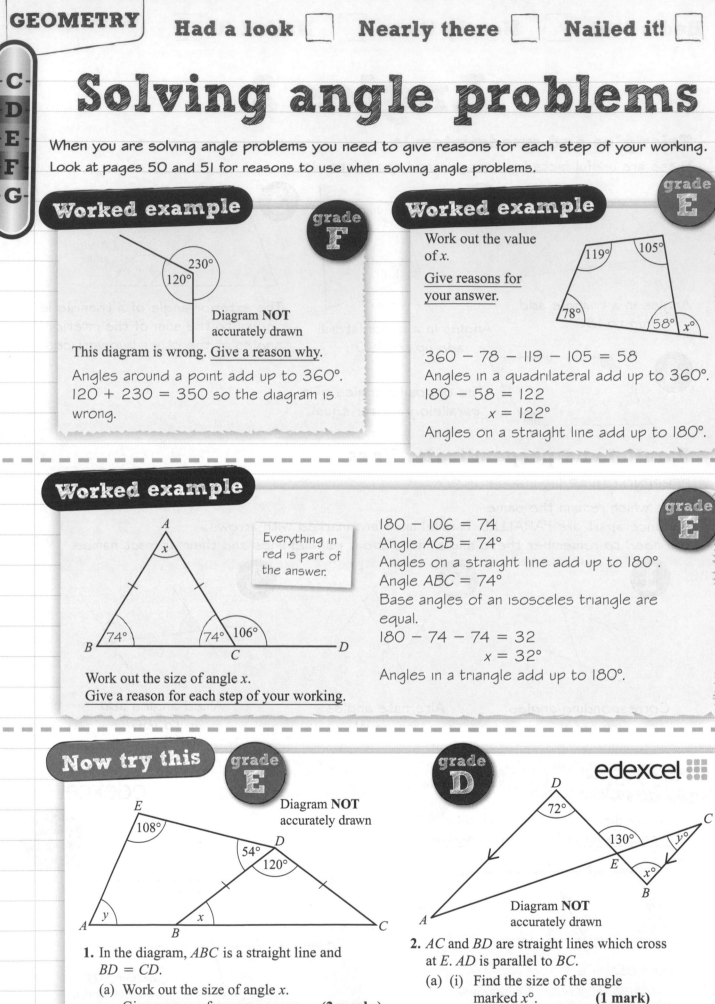

Worked example

grade **F**

230°
120°

Diagram **NOT**
accurately drawn

This diagram is wrong. Give a reason why.

Angles around a point add up to 360°.
120 + 230 = 350 so the diagram is wrong.

Worked example

grade **E**

Work out the value of x.

Give reasons for your answer.

119°   105°
78°   58° x°

360 − 78 − 119 − 105 = 58
Angles in a quadrilateral add up to 360°.
180 − 58 = 122
   x = 122°
Angles on a straight line add up to 180°.

Worked example

grade **E**

A
x
B 74°   74° 106° D
        C

Everything in red is part of the answer.

Work out the size of angle x.
Give a reason for each step of your working.

180 − 106 = 74
Angle ACB = 74°
Angles on a straight line add up to 180°.
Angle ABC = 74°
Base angles of an isosceles triangle are equal.
180 − 74 − 74 = 32
   x = 32°
Angles in a triangle add up to 180°.

Now try this

grade **E**

E
108°
          D
   54°
   120°
y   x
A   B   C

Diagram **NOT**
accurately drawn

1. In the diagram, ABC is a straight line and BD = CD.

(a) Work out the size of angle x.
Give a reason for your answer. **(2 marks)**

(b) Work out the size of angle y.
Give a reason for your answer. **(2 marks)**

grade **D**

edexcel

D
72°
      130°
          y°
      E   C
   x°
   B
A

Diagram **NOT**
accurately drawn

2. AC and BD are straight lines which cross at E. AD is parallel to BC.

(a) (i) Find the size of the angle marked x°. **(1 mark)**

(ii) Give a reason for your answer. **(1 mark)**

(b) Find the size of the angle marked y°.
(2 marks)

52

Angles in polygons

Polygon questions are all about INTERIOR and EXTERIOR angles.

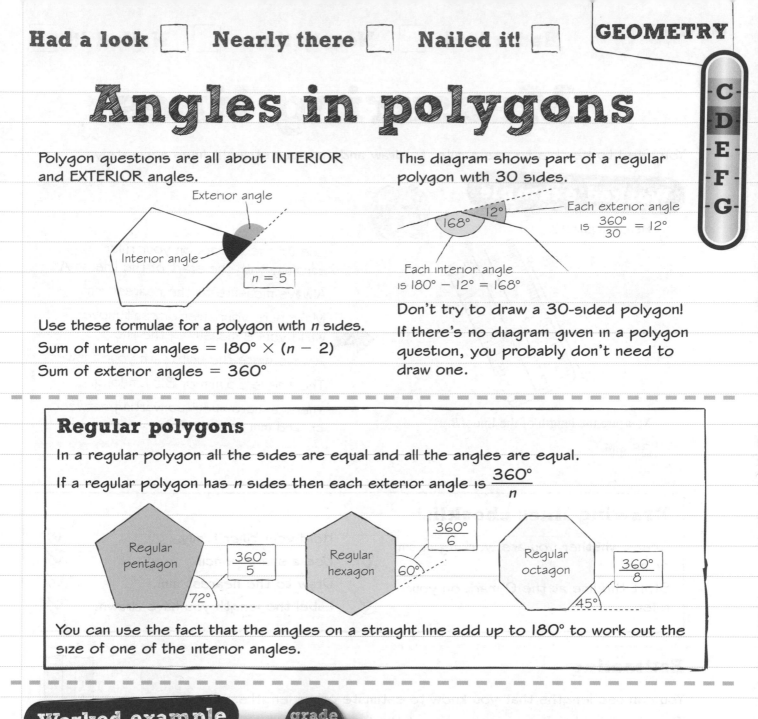

Exterior angle

Interior angle

$n = 5$

Use these formulae for a polygon with n sides.

Sum of interior angles = $180° \times (n - 2)$

Sum of exterior angles = $360°$

This diagram shows part of a regular polygon with 30 sides.

Each exterior angle is $\frac{360°}{30} = 12°$

$168°$ $12°$

Each interior angle is $180° - 12° = 168°$

Don't try to draw a 30-sided polygon! If there's no diagram given in a polygon question, you probably don't need to draw one.

Regular polygons

In a regular polygon all the sides are equal and all the angles are equal.

If a regular polygon has n sides then each exterior angle is $\frac{360°}{n}$

Regular pentagon $\frac{360°}{5}$ $72°$

Regular hexagon $60°$ $\frac{360°}{6}$

Regular octagon $45°$ $\frac{360°}{8}$

You can use the fact that the angles on a straight line add up to 180° to work out the size of one of the interior angles.

Worked example

grade **D**

Work out the size of an exterior angle of a regular pentagon.

Exterior angles of polygon add up to 360°
So exterior angle is 360° ÷ 5 = 72°

A pentagon has 5 sides. You need to know the names of the polygons with 3 to 8 sides (triangle, quadrilateral, pentagon, hexagon, heptagon, octagon) and 10 sides (decagon) for your exam.

Now try this

In part (b), start by using the interior angle to work out the exterior angle.

edexcel

(a) The size of each exterior angle of a regular polygon is 40°.
Work out the number of sides of the regular polygon. **(2 marks)**

(b) The size of each interior angle of a regular polygon is 156°.
Work out the number of sides of the polygon. **(2 marks)**

(c) The diagram shows part of a **regular** 10-sided polygon.
Work out the size of the angle marked x. **(3 marks)**

grade **C**

x

Diagram **NOT** accurately drawn

C
D
E
F
G

Measuring lines

You need to be able to use a ruler to draw and measure straight lines accurately.

Worked example

grade **G**

Here is a line *AB*.

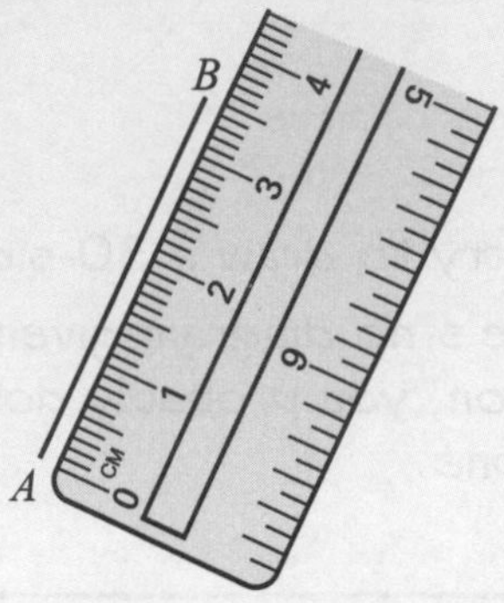

Measure the length of the line *AB*.

35 mm

Line up the 0 mark on your ruler carefully with the start of the line at A.

Always measure to the nearest mm.

Make sure your ruler doesn't move while you're measuring the line.

Always write the units with your answer.

This line is 35 mm or 3.5 cm long.

There is more about converting between cm and mm on page 62.

Drawing lines checklist

Check whether you are working in cm or mm. ✓

Start the line at the 0 mark on your ruler. ✓

Hold your ruler firmly. ✓

Use a sharp pencil. ✓

Draw to the nearest mm. ✓

Label the length you have drawn. ✓

Estimating

You can use lengths that you know to estimate other lengths.

This diagram shows a man standing at the bottom of a cliff.

The man is 3 cm tall in the drawing and the cliff is 12 cm tall.

This means the cliff is 4 times as tall as the man.

A good estimate for the height of an adult male is 1.8 m.

$4 \times 1.8 = 7.2$

This means a good estimate for the height of the cliff is 7.2 m.

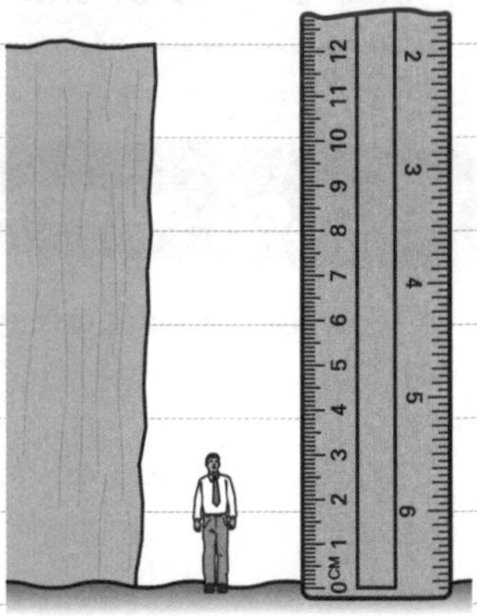

Now try this

grade **F**

edexcel

The diagram shows a man standing next to a lamp post.

The man is of normal height.

(a) Write down an estimate for the height, in metres, of the man.
(1 mark)

(b) Estimate the height, in metres, of the lamp post. **(2 marks)**

Measure the height of the man. Then see how many of these heights will fit the height of the lamp post. You only need an **estimate**.

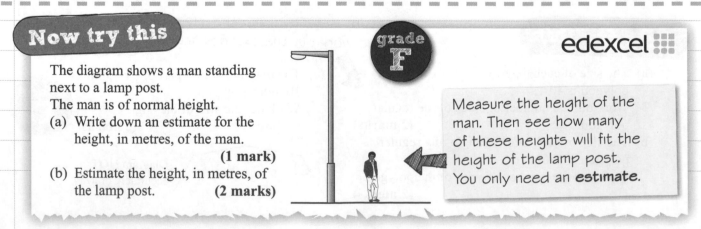

C
D
E
F
G

Bearings

BEARINGS are always measured clockwise from North.

Bearings always have 3 figures. You need to write zeros if the angle is less than 100°.

You can measure a bearing bigger than 180° by measuring this angle and subtracting it from 360°.

360° − 109° = 251°
The bearing of C from A is 251°.

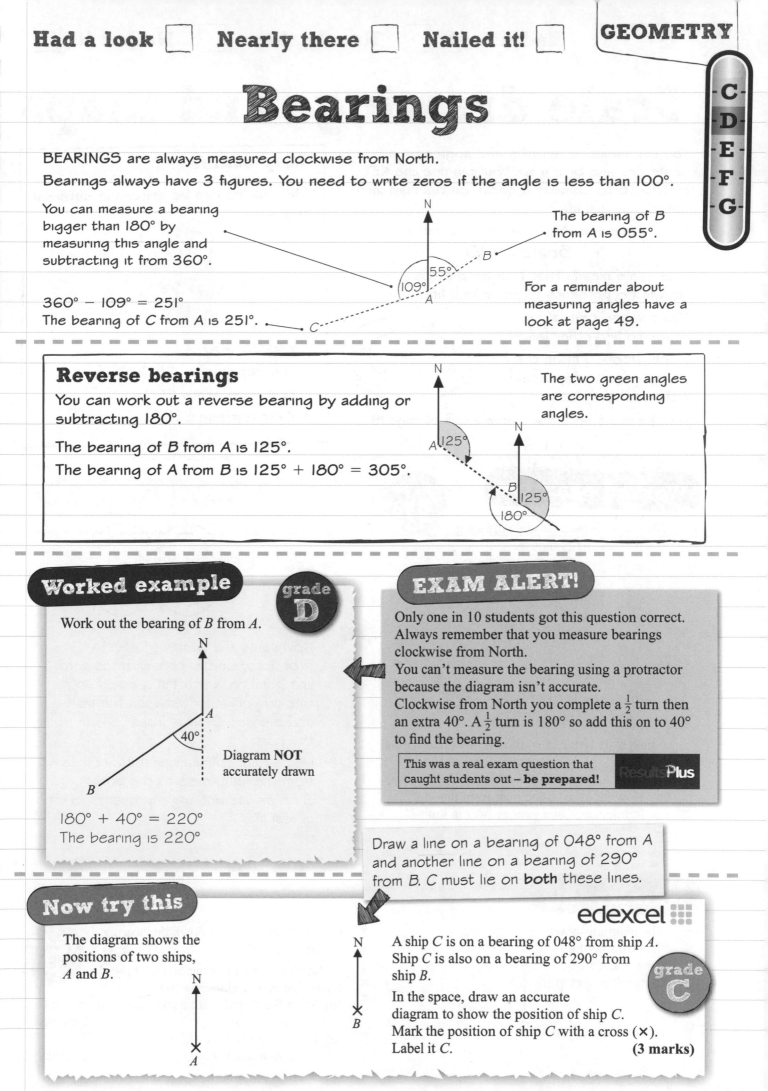

The bearing of B from A is 055°.

For a reminder about measuring angles have a look at page 49.

Reverse bearings

You can work out a reverse bearing by adding or subtracting 180°.

The bearing of B from A is 125°.

The bearing of A from B is 125° + 180° = 305°.

The two green angles are corresponding angles.

Worked example

grade D

Work out the bearing of B from A.

Diagram **NOT** accurately drawn

180° + 40° = 220°
The bearing is 220°

EXAM ALERT!

Only one in 10 students got this question correct. Always remember that you measure bearings clockwise from North.

You can't measure the bearing using a protractor because the diagram isn't accurate.

Clockwise from North you complete a $\frac{1}{2}$ turn then an extra 40°. A $\frac{1}{2}$ turn is 180° so add this on to 40° to find the bearing.

This was a real exam question that caught students out – **be prepared!**   ResultsPlus

Draw a line on a bearing of 048° from A and another line on a bearing of 290° from B. C must lie on **both** these lines.

Now try this

The diagram shows the positions of two ships, A and B.

A ship C is on a bearing of 048° from ship A. Ship C is also on a bearing of 290° from ship B.

In the space, draw an accurate diagram to show the position of ship C. Mark the position of ship C with a cross (✗). Label it C.

grade C

edexcel

(3 marks)

Scale drawings and maps

You can measure lengths and angles on a scale drawing or a map. You can use the KEY or SCALE to convert lengths on the map into lengths in real life.

Scales are sometimes given as a ratio.

Scale 1 : 10

This scale means that 1 cm on the scale drawing represents 10 cm in real life.

You can use equivalent ratios to find other lengths.

A length of 6 cm on the drawing would represent 60 cm in real life.

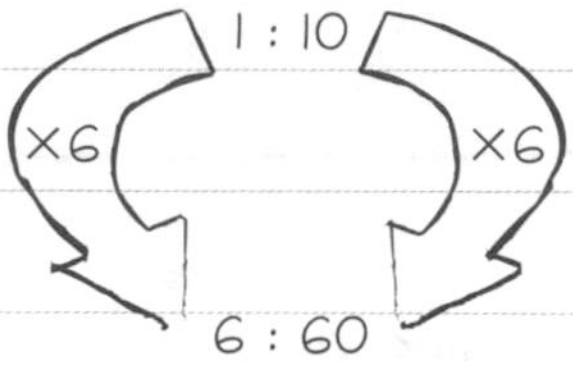

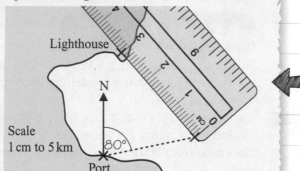

For a reminder about ratio have a look at page 19.

Map scales

Map scales can be written in different ways.

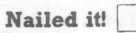

MAP

Scale
1 : 25 000

• 1 to 25 000
• 1 cm represents 25 000 cm
• 1 cm represents 250 m
• 4 cm represent 1 km

Worked example

grade D

The diagram shows a scale drawing of a port and a lighthouse.

Lighthouse

N

Scale
1 cm to 5 km   80°

Port

A boat sails 12 km in a straight line on a bearing of 080°.

(a) Mark the new position of the boat with a cross.
(b) How far away is the boat from the lighthouse? Give your answer in km.

15 km

(a) Start by working out how far the boat is from the port on the scale drawing.

÷5   ×12

Map	1 cm	0.2 cm	2.4 cm
Real life	5 km	1 km	12 km

÷5   ×12

Now place the centre of your protractor on the port with the zero line pointing North. Put a dot at 80°. Line up your ruler between the port and the dot. Draw a cross 2.4 cm from the port.

(b) Use a ruler to measure the distance from the lighthouse to the boat. 3 cm on the drawing represents 15 km in real life.

Now try this

edexcel

The diagram shows the position of a farm F and a bridge B on a map.

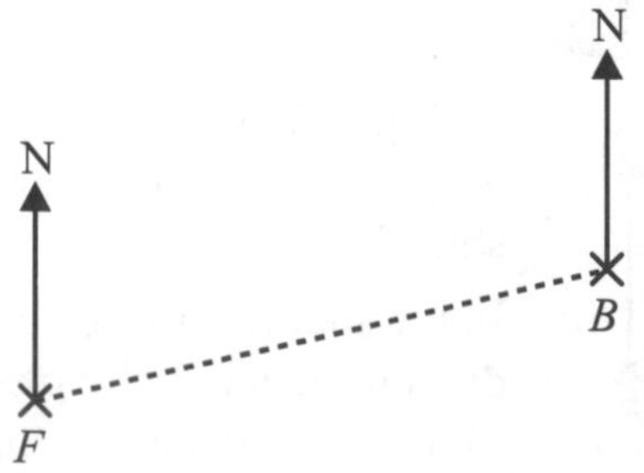

N

N

B

Scale: 1 cm = 4 km   F

(a) Measure and write down the bearing of B from F.   **(1 mark)**

grade D

A church C is on a bearing of 245° from the bridge B. The church is 10 km from B.
(b) Mark the church with a cross (✗) and label it C.   **(2 marks)**
(c) How far away is church C from farm F? Give your answer in km.   **(2 marks)**

Symmetry

Lines of symmetry

A line of symmetry is a mirror line. One half of the shape is a mirror image of the other.

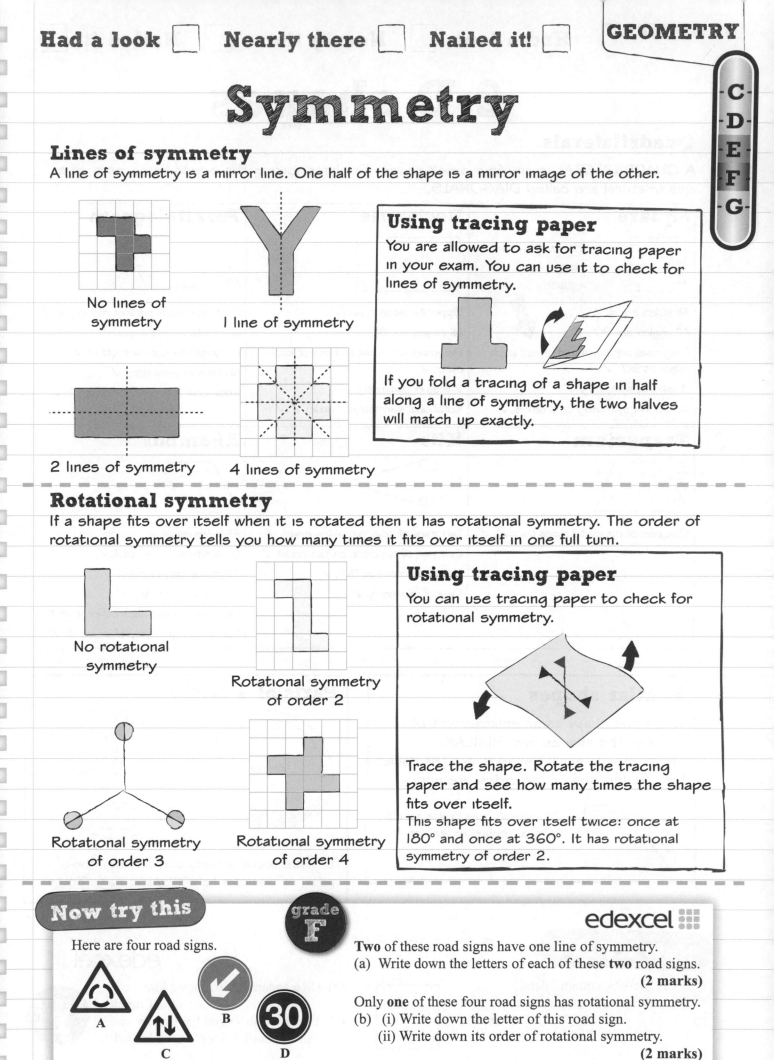

No lines of symmetry

1 line of symmetry

Using tracing paper

You are allowed to ask for tracing paper in your exam. You can use it to check for lines of symmetry.

If you fold a tracing of a shape in half along a line of symmetry, the two halves will match up exactly.

2 lines of symmetry

4 lines of symmetry

Rotational symmetry

If a shape fits over itself when it is rotated then it has rotational symmetry. The order of rotational symmetry tells you how many times it fits over itself in one full turn.

No rotational symmetry

Rotational symmetry of order 2

Rotational symmetry of order 3

Rotational symmetry of order 4

Using tracing paper

You can use tracing paper to check for rotational symmetry.

Trace the shape. Rotate the tracing paper and see how many times the shape fits over itself.

This shape fits over itself twice: once at 180° and once at 360°. It has rotational symmetry of order 2.

Now try this

grade F

edexcel

Here are four road signs.

A   B   C   D   30

Two of these road signs have one line of symmetry.
(a) Write down the letters of each of these **two** road signs.
    (2 marks)

Only **one** of these four road signs has rotational symmetry.
(b) (i) Write down the letter of this road sign.
    (ii) Write down its order of rotational symmetry.
    (2 marks)

2-D shapes

Quadrilaterals

A QUADRILATERAL is any four-sided shape. The lines joining the opposite corners of a quadrilateral are called DIAGONALS.

Square	Rectangle	Parallelogram
All sides equal. ✓	Opposite sides equal. ✓	Opposite sides parallel and equal. ✓
All angles are 90°. ✓	All angles are 90°. ✓	Opposite angles equal. ✓
Diagonals are equal and bisect each other at 90°. ✓	Diagonals are equal and bisect each other. ✓	Diagonals bisect each other. ✓
4 lines of symmetry. ✓	2 lines of symmetry. ✓	No lines of symmetry. ✓
Rotational symmetry of order 4. ✓	Rotational symmetry of order 2. ✓	Rotational symmetry of order 2. ✓

> **Bisect** means 'cut in half exactly'.

Trapezium	Kite	Rhombus
One pair of opposite sides parallel. ✓	Two pairs of adjacent sides equal. ✓	All sides equal. ✓
	One pair of opposite angles equal. ✓	Opposite sides parallel. ✓
	Diagonals cross at 90°. ✓	Opposite angles equal. ✓
	1 line of symmetry. ✓	2 lines of symmetry. ✓
		Diagonals bisect each other at 90°. ✓
		Rotational symmetry of order 2. ✓

Similar shapes

When one shape is an enlargement of another the shapes are SIMILAR.
The angles in similar shapes are the same, but the lengths of the sides are not.

Shapes A and B are similar. B is an enlargement of A.

Parts of a circle

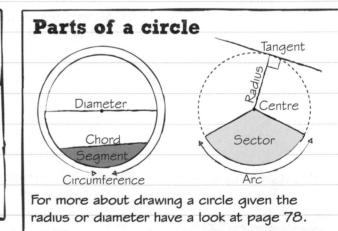

For more about drawing a circle given the radius or diameter have a look at page 78.

For more about drawing a circle given the radius or diameter have a look at page 78.

Now try this

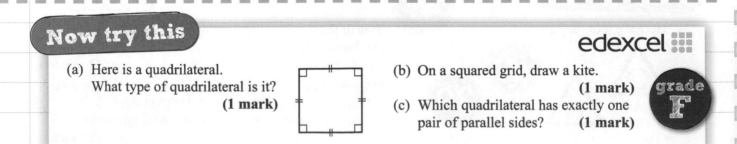

(a) Here is a quadrilateral.
   What type of quadrilateral is it?
   (1 mark)

(b) On a squared grid, draw a kite.
   (1 mark)

(c) Which quadrilateral has exactly one pair of parallel sides?   **(1 mark)**

edexcel

grade
F

C
D
E
F
G

Congruent shapes

CONGRUENT shapes are exactly the same shape and size.

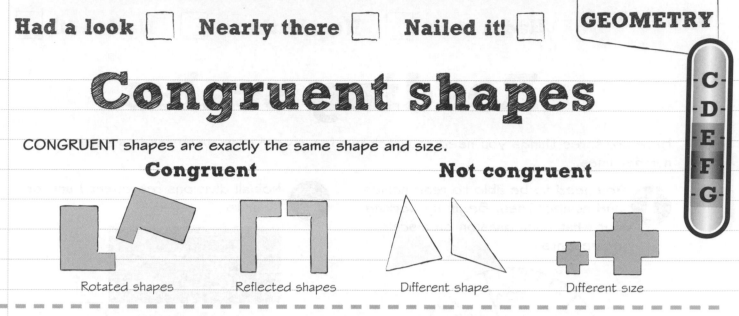

Congruent   **Not congruent**

Rotated shapes   Reflected shapes   Different shape   Different size

---

Shapes which TESSELLATE can cover a flat surface with no gaps.

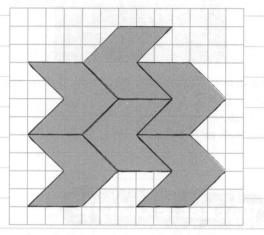

You can rotate the shape to make it fit.
All the shapes in this tessellation are congruent.

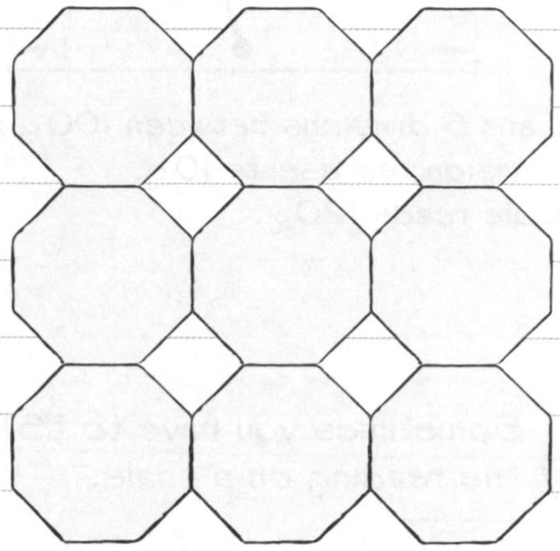

This octagon doesn't tessellate. You wouldn't be able to cover a floor using only tiles of this shape – there would be gaps!

Worked example

grade **G**

The diagram shows six triangles.
Write down the letters of **two pairs** of congruent triangles.

A and C
D and B

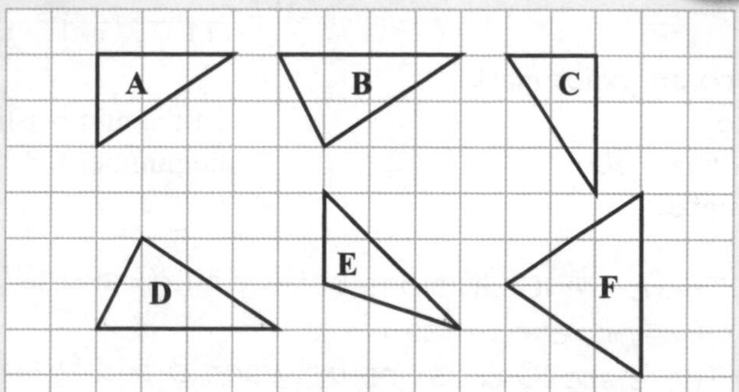

The easiest way to answer a congruency question in your exam is to ask for tracing paper. If one shape fits over a tracing of another then they are congruent. You can rotate the tracing paper or turn it upside down.

Now try this

edexcel

On the grid, show how the shaded shape will tessellate.
You must draw at least 6 shapes. **(2 marks)**

grade **E**

Remember that you can rotate the shape.

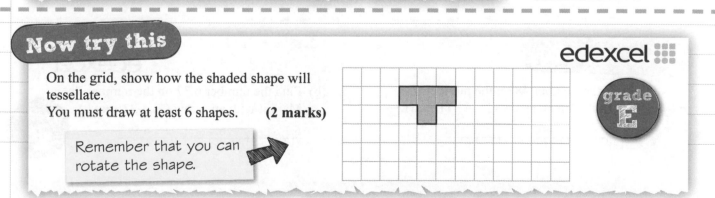

59

Had a look ☐ Nearly there ☐ Nailed it! ☐

Reading scales

Here are three things you need to watch out for in your exam when reading scales and number lines.

1 You need to be able to read scales and number lines. Begin by working out what each division on a scale represents.

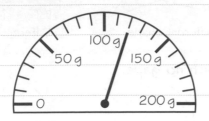

There are 5 divisions between 100 g and 150 g.
Each division represents 10 g.
The scale reads 120 g.

2 Not all divisions represent 1 unit or 10 units.

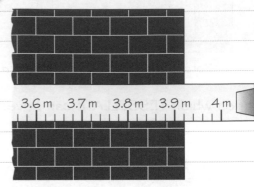

3.6 m   3.7 m   3.8 m   3.9 m   4 m

There are 5 divisions between 3.9 m and 4 m.
0.1 ÷ 5 = 0.02
Each division represents 0.02 m.
This wall is 3.92 m long.

3 Sometimes you have to ESTIMATE the reading on a scale.

50
40
30
20
10
0 ml

The water doesn't come up to an exact mark but you can make an estimate.
The water is closer to 30 ml than 40 ml.
32 ml would be a good estimate.

Worked example

grade **G**

(a) 110   120   130

Write down the number marked with the arrow.
124

grade **F**

(b) 7   8   9

On the number line mark with an arrow (↑) the number 8.8

(a) There are 10 divisions between 120 and 130 so each division represents 1
(b) There are 10 divisions between 8 and 9 so each division represents 0.1

Now try this

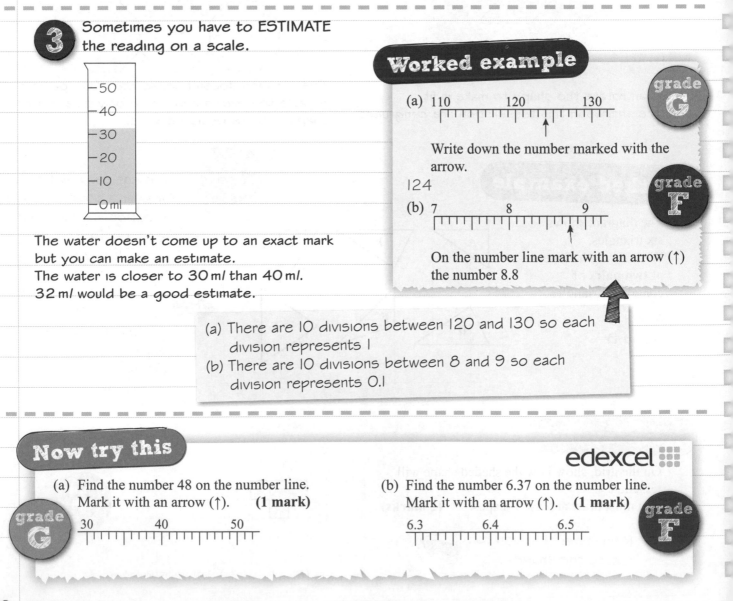

edexcel ▦

(a) Find the number 48 on the number line.
Mark it with an arrow (↑). **(1 mark)**

grade **G**

30   40   50

(b) Find the number 6.37 on the number line.
Mark it with an arrow (↑). **(1 mark)**

grade **F**

6.3   6.4   6.5

Time and timetables

-C
-D
-E
-F
-G

Time

1 You can write time using the 12-hour clock or the 24-hour clock.

12-hour clock	24-hour clock
8.15 am	08:15
4.50 pm	16:50
12.00 midday	12:00
12.00 midnight	00:00

2 Remember there are 60 minutes in an hour.

Hours	Minutes
$\frac{1}{2}$	30
$\frac{3}{4}$	45
$2\frac{1}{4}$	135

You can also write this as '2 hours and 15 minutes'.
60 + 60 + 15 = 135

Timetables

This table shows part of a bus timetable.

Timetables usually give times using the 24-hour clock.

Crook	08:15	09:15	10:45	11:15
Prudhoe	08:28	09:28	10:58	11:28
Hexham	08:45	09:45	11:15	11:45
Alton	09:00	10:00	11:30	12:00

This bus leaves Crook at 10:45 and arrives in Hexham at 11:15

This bus leaves Prudhoe at 11:28 and arrives in Alton at 12:00
11:28 to 11:30 = 2 minutes
11:30 to 12:00 = 30 minutes
The journey time is 32 minutes

Worked example

grade **F**

Olivia is visiting her sister in Exeter. She leaves home at 11.30 am. She writes down how long each section of her journey takes.

Drive to train station   15 minutes
Wait for train   25 minutes
Train journey $1\frac{1}{2}$ hours

At what time does Olivia arrive in Exeter?

11.30 am → 11.45 am → 12.10 pm → 1.40 pm
Olivia arrives at 1.40 pm.

Add on each time in steps. Write down each new time.

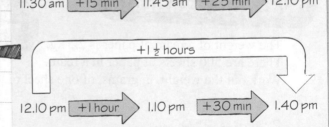

Now try this

edexcel

Here is part of a railway timetable.

Bath	08:15	08:45	09:15
Chippenham	08:30	09:00	09:30
Swindon	08:50	09:20	09:50
Didcot	09:15	09:45	10:15
Reading	09:35	10:05	10:35

Jamelia gets to the station in Chippenham at 08:45
She waits for the next train to Didcot.

(a) (i) How long should she have to wait?
    (ii) At what time should she arrive at Didcot? **(2 marks)**

All the trains should take the same time to travel from Bath to Reading.

(b) How long, in minutes, should it take to travel from Bath to Reading? **(2 marks)**

grade **F**

Metric units

Most of the units of measurement used in the UK are METRIC units.

You can convert between metric units by multiplying or dividing by 10, 100 or 1000.

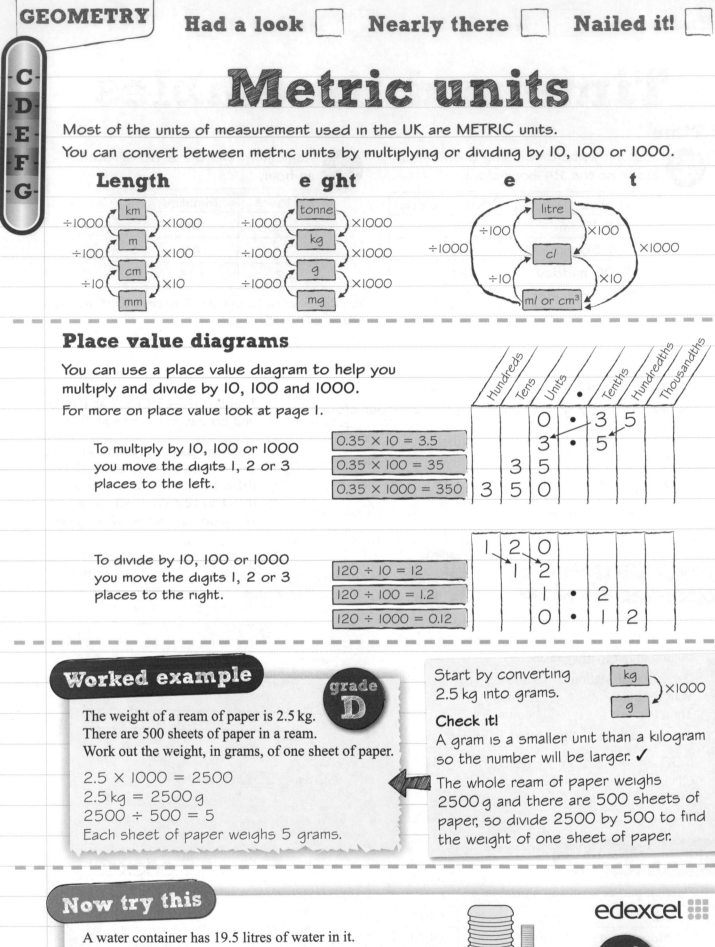

Length

```
÷1000 ⤵ km ⤴ ×1000
÷100  ⤵ m  ⤴ ×100
÷10   ⤵ cm ⤴ ×10
         mm
```

e ght

```
÷1000 ⤵ tonne ⤴ ×1000
÷1000 ⤵ kg    ⤴ ×1000
÷1000 ⤵ g     ⤴ ×1000
         mg
```

e   t

```
÷1000        litre        ×1000
        ÷100 ⤵   ⤴ ×100
              cl
        ÷10 ⤵   ⤴ ×10
          ml or cm³
```

Place value diagrams

You can use a place value diagram to help you multiply and divide by 10, 100 and 1000.

For more on place value look at page 1.

To multiply by 10, 100 or 1000 you move the digits 1, 2 or 3 places to the left.

| $0.35 \times 10 = 3.5$ |
| $0.35 \times 100 = 35$ |
| $0.35 \times 1000 = 350$ |

Hundreds	Tens	Units	•	Tenths	Hundredths	Thousandths
		0	•	3	5	
		3	•	5		
	3	5				
3	5	0				

To divide by 10, 100 or 1000 you move the digits 1, 2 or 3 places to the right.

| $120 \div 10 = 12$ |
| $120 \div 100 = 1.2$ |
| $120 \div 1000 = 0.12$ |

			•			
1	2	0				
	1	2				
		1	•	2		
		0	•	1	2	

The weight of a ream of paper is 2.5 kg.
There are 500 sheets of paper in a ream.
Work out the weight, in grams, of one sheet of paper.

$2.5 \times 1000 = 2500$
$2.5 \, kg = 2500 \, g$
$2500 \div 500 = 5$
Each sheet of paper weighs 5 grams.

Start by converting 2.5 kg into grams.

```
kg ⤵ ×1000
g
```

Check it!
A gram is a smaller unit than a kilogram so the number will be larger. ✓

The whole ream of paper weighs 2500 g and there are 500 sheets of paper, so divide 2500 by 500 to find the weight of one sheet of paper.

Now try this

edexcel

A water container has 19.5 litres of water in it.
A cup holds 210 ml of water.
At most 92 cups can be filled completely from the water container.
Explain why.
You must show all your working.   **(3 marks)**

grade **D**

Measures

Most of the units of measurement used in the UK are METRIC units like litres, kilograms and kilometres. You can convert these into IMPERIAL units like gallons, pounds and miles.

A question may ask you to choose a sensible unit for a measurement.

A sensible unit for the distance from Liverpool to Bath is km (metric) or miles (imperial).

For a reminder about questions on estimating measures and reading scales have a look at pages 54 and 60.

Metric and imperial

You need to remember these conversions for your exam.

Metric unit	Imperial unit
1 kg	2.2 pounds (lb)
1 litre (l)	$1\frac{3}{4}$ pints
4.5 litres	1 gallon
8 km	5 miles
30 cm	1 foot (ft)

When converting between different imperial units you will be GIVEN the conversions.

For a reminder about metric units have a look at page 62.

Worked example   grade D

The UK speed limit in most towns is 30 mph. In France the speed limit in towns is 50 km/h. Is the speed limit higher in the UK or in France? You must show all your working.

8 km = 5 miles
UK: 30 ÷ 5 = 6
     8 × 6 = 48
     30 mph = 48 km/h
48 km/h is less than 50 km/h so the speed limit is higher in France.

To compare the speeds both speed limits need to be in the same units. Convert 30 miles into km. You can use equivalent ratios:

Remember to compare the two speed limits and write your answer as a sentence.

miles : km
5 : 8   (×6)
30 : 48   (×6)

Speed

To convert between measures of speed using metric units you need to convert one unit first then the other.

72 km/h → m/s
72 km/h → 72 × 1000 = 72 000 m/h
72 000 m/h → 72 000 ÷ 3600 = 20 m/s

1 hour = 60 × 60 = 3600 seconds

Volume and capacity

You can convert between metric units of volume and units of capacity using:

1 cm³ = 1 ml
1000 cm³ = 1 litre
1 m³ = 1000 litres

There is more on volume conversions on page 75.

Now try this

Look at page 64 for more on speed.

edexcel

1. The motorway speed limit in Great Britain is 70 mph.

   The motorway speed limit in Spain is 120 km/h.

   Which of these speed limits is lower? You must show all your working. **(3 marks)**

   grade D

2. Jane walks at an average speed of 5 km/h.

   Mattie walks at an average speed of 3 miles per hour.

   How long will they each take to walk 5 miles? **(3 marks)**

   grade C

C
D
E
F
G

Speed

This is the FORMULA TRIANGLE for SPEED.

Distance

Average speed

Time

$$\text{Average speed} = \frac{\text{total distance travelled}}{\text{total time taken}}$$

$$\text{Time} = \frac{\text{distance}}{\text{average speed}}$$

$$\text{Distance} = \text{average speed} \times \text{time}$$

Using a formula triangle

Cover up the quantity you want to find with your finger.

The position of the other two quantities tells you the formula.

$$T = \frac{D}{S} \qquad S = \frac{D}{T} \qquad D = S \times T$$

Units

The most common units of speed are
- metres per second: m/s
- kilometres per hour: km/h
- miles per hour: mph

The units in your answer will depend on the units you use in the formula.

When distance is measured in km and time is measured in hours, speed will be measured in km/h.

Minutes and hours

For questions on speed, you need to be able to convert between minutes and hours. Look at page 61 for more on time.

Speed questions

Draw a formula triangle. ✓

Make sure the units match. ✓

Give units with your answer. ✓

Worked example

grade C

A plane travels at a constant speed of 600 km/h for 45 minutes.
How far has it travelled?

$$45 \text{ minutes} = \frac{45}{60} \text{ hours} = \frac{3}{4} \text{ hour}$$

$$D = S \times T$$
$$= 600 \times \frac{3}{4} = \frac{600 \times 3}{4} = \frac{1800}{4} = 450$$

The plane has travelled 450 km.

When you are calculating a distance, you **must** make sure that the units of the other quantities match.

Speed is given in km/h and time is given in minutes. You need to convert the time into hours before you use the formula.

(a) Change your time in hours and minutes to 2.5 hours.

(b) Use your formula for distance, speed and time but instead of *distance* use *volume* and instead of *speed* use *rate*.

Now try this

 grade C

Sally drives a distance of 100 miles.
It takes 2 hours 30 minutes for Sally to drive this distance.

(a) Work out her average speed. **(3 marks)**

edexcel

At her work Sally has to empty a tank of water.
There are 80 litres of water in the tank.
Sally turns a tap on. Water flows from the tank at a rate of 5 litres per minute.

(b) How many minutes will it take for the tank to become completely empty? **(2 marks)**

C
D
E
F
G

Perimeter and area

Perimeter

PERIMETER is the distance around the edge of a shape. You can work out the perimeter of a shape by adding up the lengths of the sides.

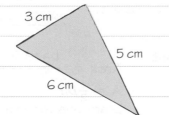

3 cm

5 cm

6 cm

Perimeter = 3 cm + 5 cm + 6 cm

= 14 cm

You might need to measure a shape to find the perimeter.

See page 54 for help on measuring lines.

Worked example

grade
E

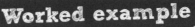
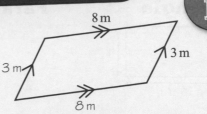

8 m

3 m

3 m

8 m

Work out the perimeter of this parallelogram.

3 + 8 + 3 + 8 = 22

Perimeter = 22 m

Work out the missing lengths first. The opposite sides of a parallelogram are equal so you can fill in these lengths on the diagram.

Area

You can work out the AREA of a shape drawn on squared paper by counting the squares.

This area is 1 cm²
You say 'one centimetre squared'
or
'one square centimetre'

Area = 9 cm²

Estimating

You might need to estimate the area of a shape drawn on cm squared paper. Count 1 cm² for every whole square and $\frac{1}{2}$ cm² for every part square.

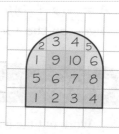

	2	3	4	5
1	9	10	6
5	6	7	8
	1	2	3	4

Here there are 10 whole squares and 6 part squares. A good estimate is 13 cm².

Worked example

grade
G

This shape is drawn on cm squared paper.

(a) Work out the perimeter of the shape.

18 cm

(b) Work out the area of the shape.

12 cm²

Golden rule

Always give units with your answer.

Units of perimeter are mm, cm, m or km.

Units of area are mm², cm², m² or km².

Now try this

edexcel

This shape is drawn on a grid of cm squares.

(a) Find the perimeter of the shape.   **(2 marks)**

(b) Find the area of the shape.   **(1 mark)**

grade
G

C
D
E
F
G

Using area formulae

You can use these formulae to work out the areas of some 2-D shapes.

Rectangle

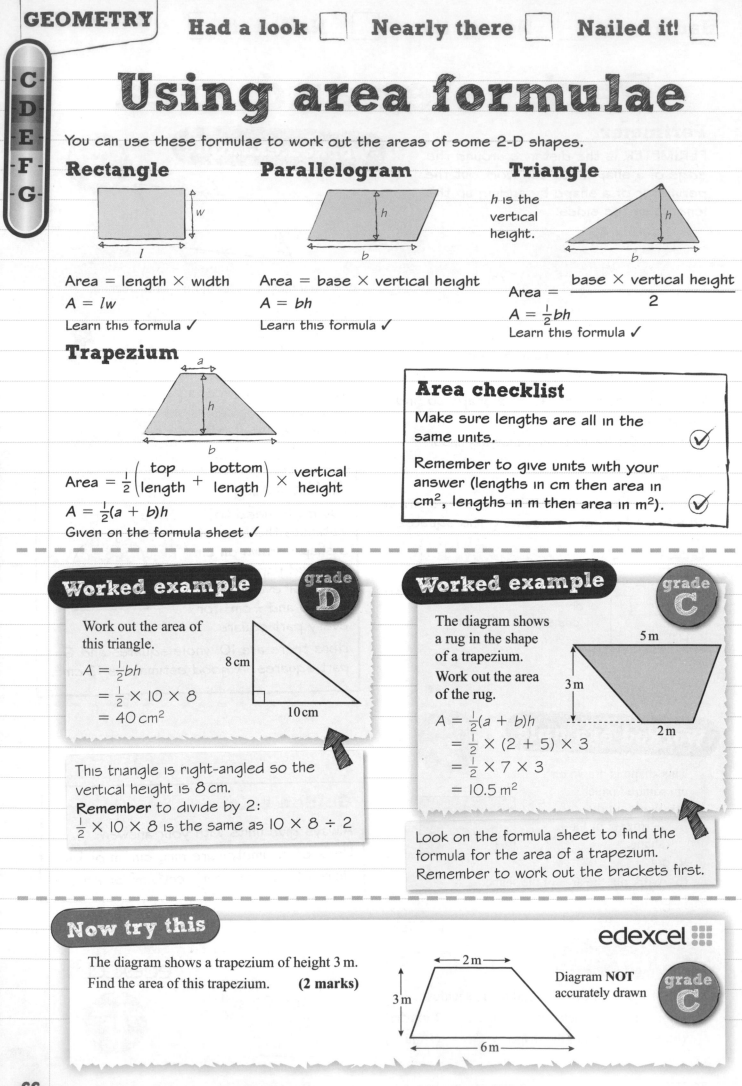

Area = length × width

$A = lw$

Learn this formula ✓

Parallelogram

Area = base × vertical height

$A = bh$

Learn this formula ✓

Triangle

h is the vertical height.

Area = $\dfrac{\text{base} \times \text{vertical height}}{2}$

$A = \frac{1}{2}bh$

Learn this formula ✓

Trapezium

Area = $\frac{1}{2}\left(\dfrac{\text{top}}{\text{length}} + \dfrac{\text{bottom}}{\text{length}}\right) \times \dfrac{\text{vertical}}{\text{height}}$

$A = \frac{1}{2}(a + b)h$

Given on the formula sheet ✓

Area checklist

Make sure lengths are all in the same units.   ✓

Remember to give units with your answer (lengths in cm then area in cm², lengths in m then area in m²).   ✓

Worked example   grade D

Work out the area of this triangle.

$A = \frac{1}{2}bh$

$= \frac{1}{2} \times 10 \times 8$

$= 40 \text{ cm}^2$

8 cm

10 cm

This triangle is right-angled so the vertical height is 8 cm.

Remember to divide by 2:

$\frac{1}{2} \times 8$ is the same as 10 × 8 ÷ 2

Worked example   grade C

The diagram shows a rug in the shape of a trapezium. Work out the area of the rug.

5 m

3 m

2 m

$A = \frac{1}{2}(a + b)h$

$= \frac{1}{2} \times (2 + 5) \times 3$

$= \frac{1}{2} \times 7 \times 3$

$= 10.5 \text{ m}^2$

Look on the formula sheet to find the formula for the area of a trapezium. Remember to work out the brackets first.

Now try this

The diagram shows a trapezium of height 3 m.
Find the area of this trapezium.   **(2 marks)**

2 m

3 m

6 m

Diagram **NOT** accurately drawn

grade C

edexcel

Solving area problems

C D E F G

You can calculate areas and perimeters of harder shapes by splitting them into parts.
You might need to draw some extra lines on your diagram and add or subtract areas.

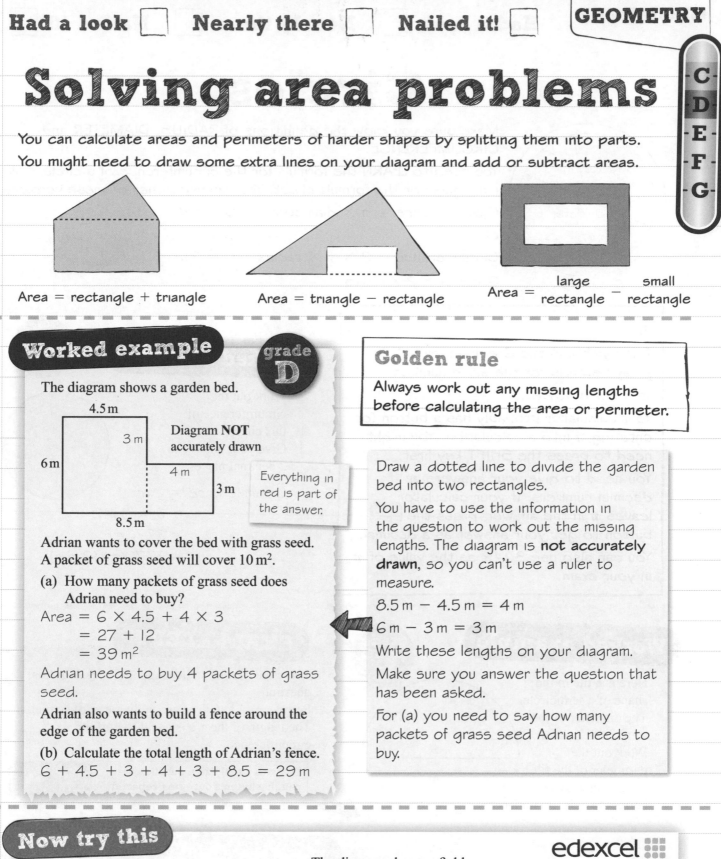

Area = rectangle + triangle   Area = triangle − rectangle   Area = large rectangle − small rectangle

Worked example

grade **D**

The diagram shows a garden bed.

4.5 m
3 m
6 m
4 m
3 m
8.5 m

Diagram **NOT** accurately drawn

Everything in red is part of the answer.

Adrian wants to cover the bed with grass seed.
A packet of grass seed will cover 10 m².

(a) How many packets of grass seed does Adrian need to buy?

Area = 6 × 4.5 + 4 × 3
 = 27 + 12
 = 39 m²

Adrian needs to buy 4 packets of grass seed.

Adrian also wants to build a fence around the edge of the garden bed.

(b) Calculate the total length of Adrian's fence.

6 + 4.5 + 3 + 4 + 3 + 8.5 = 29 m

Golden rule

Always work out any missing lengths before calculating the area or perimeter.

Draw a dotted line to divide the garden bed into two rectangles.

You have to use the information in the question to work out the missing lengths. The diagram is **not accurately drawn**, so you can't use a ruler to measure.

8.5 m − 4.5 m = 4 m

6 m − 3 m = 3 m

Write these lengths on your diagram.

Make sure you answer the question that has been asked.

For (a) you need to say how many packets of grass seed Adrian needs to buy.

Now try this

edexcel :::

Diagram **NOT** accurately drawn

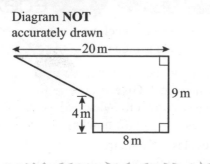

20 m
9 m
4 m
8 m

The diagram shows a field.

A farmer wants to put some sheep in the field.

He decides that each sheep needs at least 5 m² for grazing.

Work out the maximum number of sheep he can put in this field.

grade **C**

(4 marks)

1. Divide the shape into 2 parts.
2. Fill in any measurements needed for the area on the diagram.
3. Answer the question asked!

C
D
E
F
G

Circles

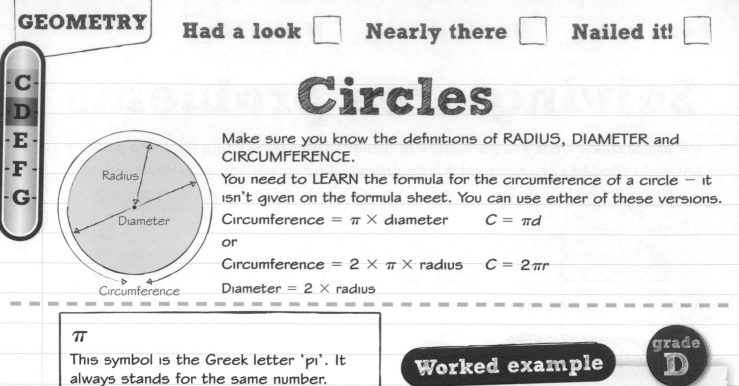

Make sure you know the definitions of RADIUS, DIAMETER and CIRCUMFERENCE.

You need to LEARN the formula for the circumference of a circle — it isn't given on the formula sheet. You can use either of these versions.

Circumference = $\pi \times$ diameter $C = \pi d$

or

Circumference = $2 \times \pi \times$ radius $C = 2\pi r$

Diameter = $2 \times$ radius

π

This symbol is the Greek letter 'pi'. It always stands for the same number.

π = 3.1415926...

Your calculator probably has a button for entering π into a calculation. You might need to press the SHIFT key first.

You need to give your answers as decimal numbers. If your calculator leaves π in the answer, press the [S⇔D] button to get your answer as a decimal.

You can also use 3.142 as the value of π in your exam.

Worked example
grade D

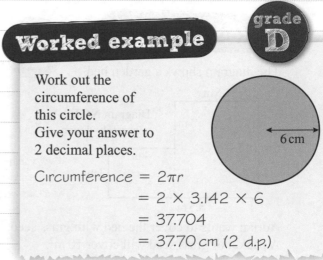

Work out the circumference of this circle. Give your answer to 2 decimal places.

Circumference = $2\pi r$

= $2 \times 3.142 \times 6$

= 37.704

= 37.70 cm (2 d.p.)

Worked example
grade C

Here is a tile in the shape of a semicircle. The diameter of the semicircle is 8 cm. Work out the perimeter of the tile. Give your answer correct to 2 decimal places.

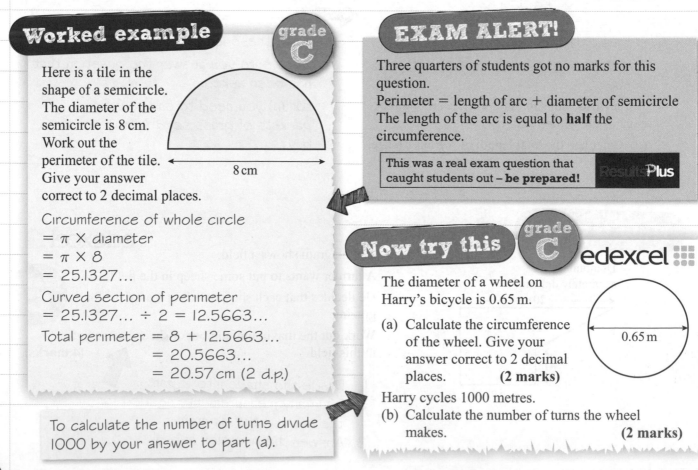

8 cm

Circumference of whole circle

= $\pi \times$ diameter

= $\pi \times 8$

= 25.1327...

Curved section of perimeter

= 25.1327... ÷ 2 = 12.5663...

Total perimeter = 8 + 12.5663...

= 20.5663...

= 20.57 cm (2 d.p.)

EXAM ALERT!

Three quarters of students got no marks for this question.

Perimeter = length of arc + diameter of semicircle

The length of the arc is equal to **half** the circumference.

> This was a real exam question that caught students out – **be prepared!**   ResultsPlus

---

Now try this
grade C   edexcel

The diameter of a wheel on Harry's bicycle is 0.65 m.

0.65 m

(a) Calculate the circumference of the wheel. Give your answer correct to 2 decimal places.   **(2 marks)**

Harry cycles 1000 metres.

(b) Calculate the number of turns the wheel makes.   **(2 marks)**

To calculate the number of turns divide 1000 by your answer to part (a).

C
D
E
F
G

Area of a circle

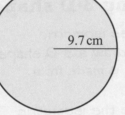

Radius

You need to LEARN the formula for the area of a circle — it isn't given on the formula sheet.

Area = π × radius²

$A = \pi \times r \times r = \pi r^2$

You always need to use the RADIUS when you are calculating the area.

If you are given the diameter, use radius = diameter ÷ 2 = $\dfrac{d}{2}$

Worked example grade D

9.7 cm

Work out the area of this circle.
Give your answer to 2 decimal places.

Area = πr^2

= π × 9.7²

= π × 94.09

= 295.5924...

= 295.59 cm² (2 d.p.)

Write down at least 4 decimal places from your calculator display before giving your final answer.

Make sure you include units with your answer. The radius is given in cm so the units of area are cm².

Worked example grade C

15 cm

The diameter of this semicircle is 15 cm.
Calculate the area of the semicircle.
Give your answer correct to 2 decimal places.

Radius = 15 ÷ 2 = 7.5 cm

Area of whole circle = πr^2

= π × 7.5²

= π × 56.25

= 176.7145... cm²

Area of semicircle = 176.7145... ÷ 2

= 88.3572...

= 88.36 cm² (2 d.p.)

A semicircle is **half** a circle.
Be careful: you are given the **diameter** of the circle and you need the **radius**.

Now try this

Work out the area of the square and work out the area of the circle. Subtract to find the difference.

edexcel

6 cm

12 cm

12 cm

A circle has a radius of 6 cm.

A square has a side of length 12 cm.

Work out the difference between the area of the circle and the area of the square.

Give your answer correct to 1 decimal place.

(4 marks)

grade C

3-D shapes

You need to learn the names of these 3-D shapes.

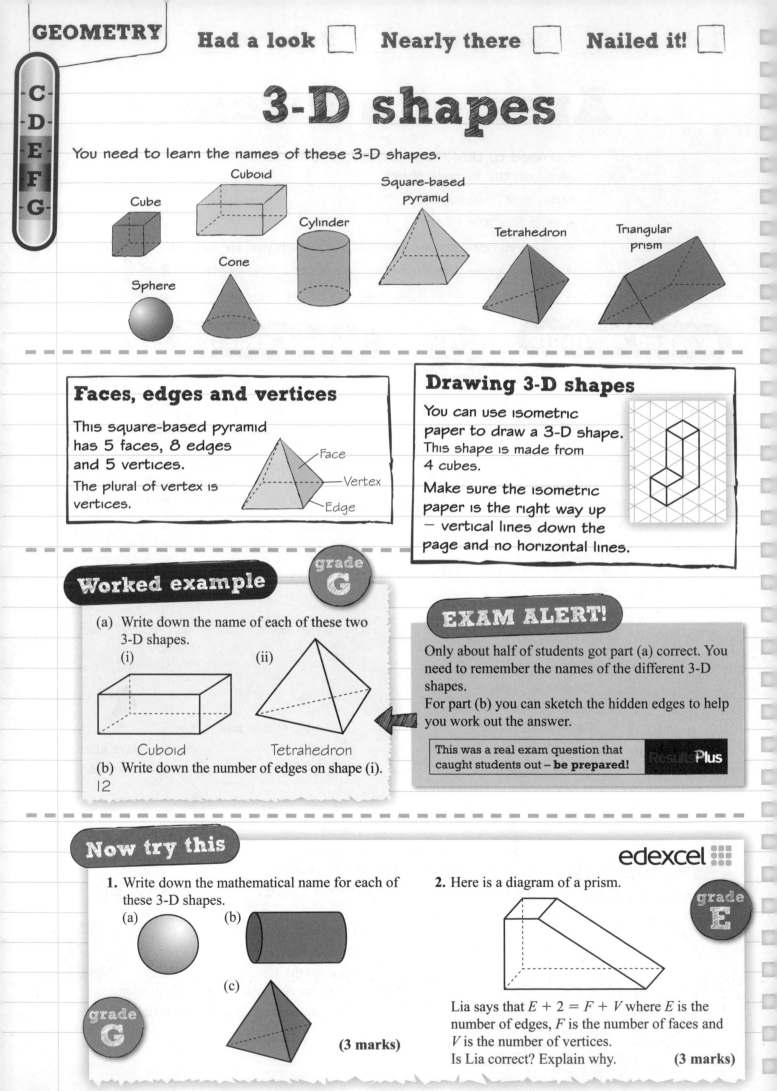

Cube   Cuboid   Cylinder   Square-based pyramid   Tetrahedron   Triangular prism   Cone   Sphere

Faces, edges and vertices

This square-based pyramid has 5 faces, 8 edges and 5 vertices.

The plural of vertex is vertices.

Face
Vertex
Edge

Drawing 3-D shapes

You can use isometric paper to draw a 3-D shape. This shape is made from 4 cubes.

Make sure the isometric paper is the right way up — vertical lines down the page and no horizontal lines.

Worked example

grade G

(a) Write down the name of each of these two 3-D shapes.
(i)   (ii)

Cuboid   Tetrahedron

(b) Write down the number of edges on shape (i).
12

EXAM ALERT!

Only about half of students got part (a) correct. You need to remember the names of the different 3-D shapes.

For part (b) you can sketch the hidden edges to help you work out the answer.

This was a real exam question that caught students out – **be prepared!**

ResultsPlus

Now try this

edexcel

1. Write down the mathematical name for each of these 3-D shapes.

(a)   (b)

(c)

(3 marks)

grade G

2. Here is a diagram of a prism.

grade E

Lia says that $E + 2 = F + V$ where E is the number of edges, F is the number of faces and V is the number of vertices.
Is Lia correct? Explain why.   **(3 marks)**

Plan and elevation

Plans and elevations are 2-D drawings of 3-D shapes as seen from different directions.

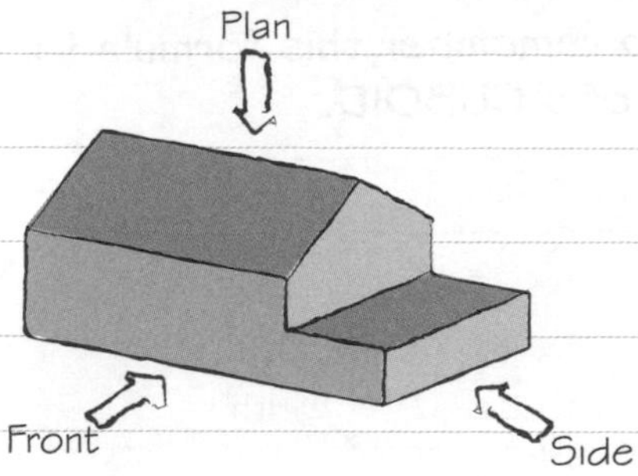

Plan

Front   Side

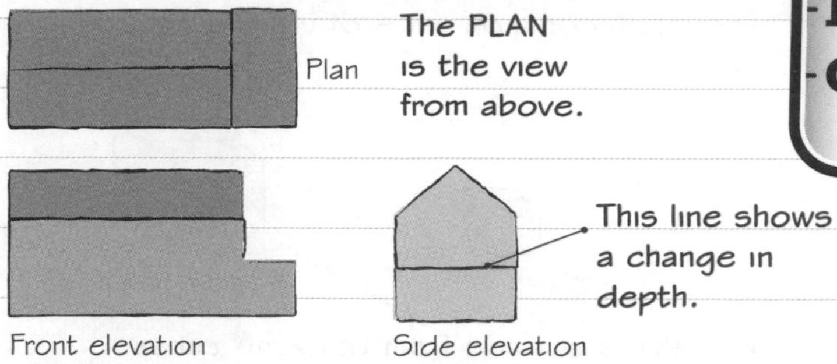

Plan

The PLAN is the view from above.

Front elevation

Side elevation

This line shows a change in depth.

The FRONT ELEVATION is the view from the front.

The SIDE ELEVATION is the view from the side.

A NET of a 3-D shape is a flat pattern that can be folded up to make the shape.
This is a net of a cube.

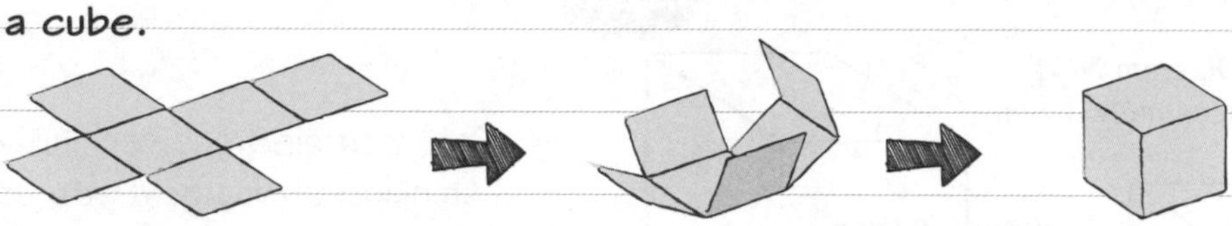

To decide if a pattern is a net of a 3-D shape imagine trying to fold it up to make the shape.

Worked example

grade D

The diagram shows a solid shape.

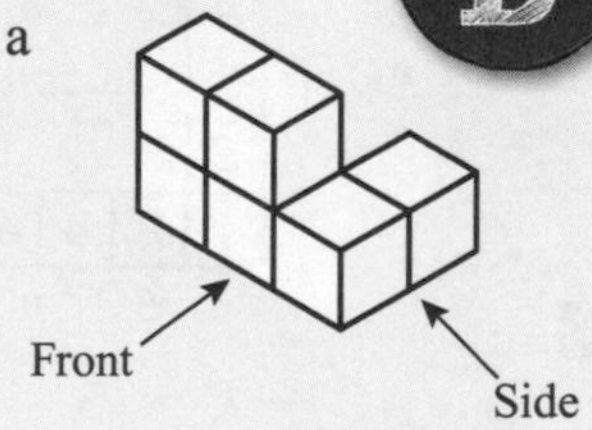

Front   Side

On the grid below draw a plan, front and side elevations of the shape.

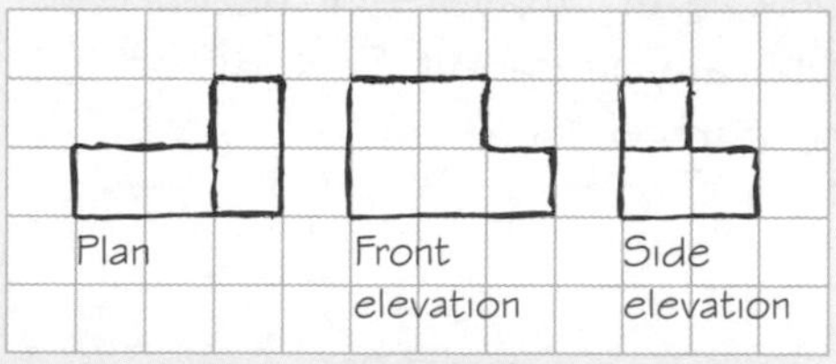

Plan   Front elevation   Side elevation

Imagine tracing an image of the shape on each side of a box.
Unfold the box to get your plan and elevations.

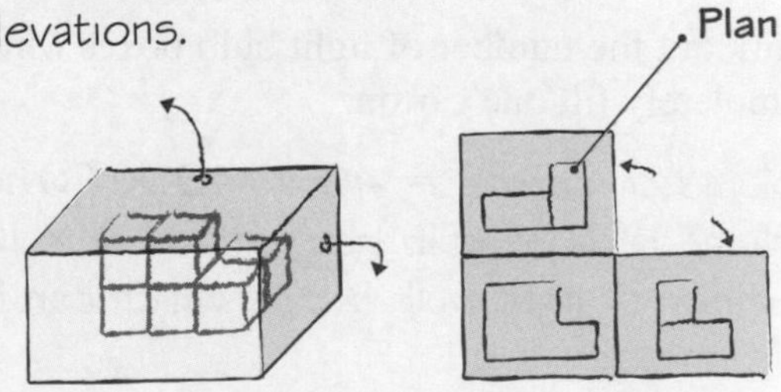

Plan

Put lines within the plan and side elevation to show where there is a change in height or depth.

The vertical lines on the plan show where the height of the shape changes.

Now try this

Here are the plan and front elevation of a prism.
The front elevation shows the cross-section of the prism.

(a) On a grid, draw a side elevation of the prism.   **(3 marks)**

(b) Draw a 3-D sketch of the prism.   **(2 marks)**

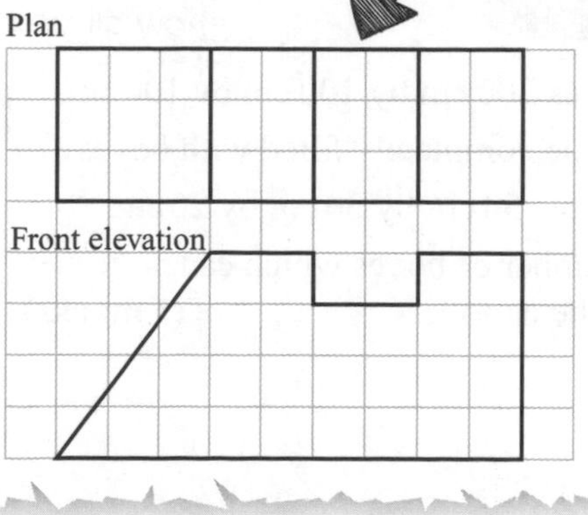

Plan

Front elevation

edexcel

grade C

C
D
E
F
G

Volume

The VOLUME of a 3-D shape is the amount of space it takes up.

The most common units of volume are cm³ or m³.

Volume = 1 cm³   Volume = 10 cm³

This shape is made from ten 1 cm³ cubes.

Volume of a cuboid

You need to remember this formula for the volume of a CUBOID.

height

width

length

Volume = length × width × height

Worked example

grade **D**

Diagram **NOT** accurately drawn

8 cm

8 cm   10 cm

40 cm   Carton

40 cm   60 cm

A light bulb box measures 8 cm by 8 cm by 10 cm.

Light bulb boxes are packed into cartons.

A carton measures 40 cm by 40 cm by 60 cm.

Work out the number of light bulb boxes which can completely fill one carton.

Volume of carton = 40 × 40 × 60 cm³
Volume of light bulb box = 8 × 8 × 10 cm³
Number of light bulb boxes which can fill one carton

$$= \frac{\overset{5}{40} \times \overset{5}{40} \times \overset{6}{60}}{\underset{1}{8} \times \underset{1}{8} \times \underset{1}{10}}$$

$$= 5 \times 5 \times 6$$
$$= 150$$

You can also answer this question by thinking about how you can fit the light bulb boxes into the carton.

40 cm   Carton   60 cm

40 cm

$40 \div 8 = 5$ $60 \div 10 = 6$

You can fit 5 × 6 = 30 boxes in each layer.
You can fit 5 layers in the carton.
In total you can fit 5 × 30 = 150 boxes in the carton.

Now try this

Because the carton is to be completely filled you can use either of the methods used in the Worked example. Be sure to show all your working.

edexcel

A carton measures 200 cm by 100 cm by 100 cm.

The carton is to be completely filled with boxes.

Each box measures 50 cm by 20 cm by 20 cm.

Work out the number of boxes which can completely fill the carton.   **(3 marks)**

grade **D**

Diagrams **NOT** accurately drawn

CARTON

200 cm   100 cm

100 cm

BOX   20 cm

50 cm   20 cm

Prisms

C
D
E
F
G

Volume

A prism is a 3-D shape with a constant CROSS-SECTION.
This formula for the VOLUME of a prism is given on your formula sheet.

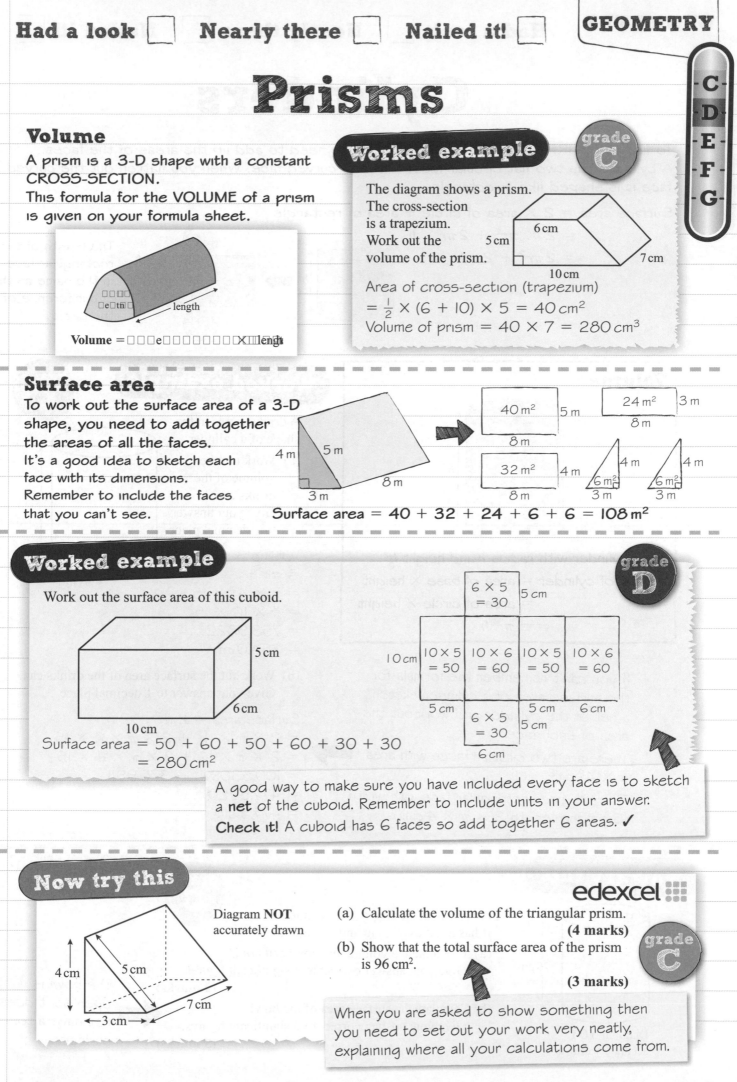

Volume = □□□e□□□□□□□□ × □ lengt□

Worked example

The diagram shows a prism.
The cross-section is a trapezium.
Work out the volume of the prism.

6 cm
5 cm
7 cm
10 cm

Area of cross-section (trapezium)
= $\frac{1}{2}$ × (6 + 10) × 5 = 40 cm²
Volume of prism = 40 × 7 = 280 cm³

Surface area

To work out the surface area of a 3-D shape, you need to add together the areas of all the faces.
It's a good idea to sketch each face with its dimensions.
Remember to include the faces that you can't see.

4 m 5 m
3 m 8 m

40 m² 5 m
8 m

24 m² 3 m
8 m

32 m² 4 m
8 m

4 m
6 m²
3 m

4 m
6 m²
3 m

Surface area = 40 + 32 + 24 + 6 + 6 = 108 m²

Worked example

Work out the surface area of this cuboid.

5 cm
6 cm
10 cm

6 × 5 = 30 5 cm

10 cm 10 × 5 = 50 | 10 × 6 = 60 | 10 × 5 = 50 | 10 × 6 = 60

5 cm 5 cm 6 cm
6 × 5 = 30 5 cm
6 cm

Surface area = 50 + 60 + 50 + 60 + 30 + 30
= 280 cm²

A good way to make sure you have included every face is to sketch a **net** of the cuboid. Remember to include units in your answer.
Check it! A cuboid has 6 faces so add together 6 areas. ✓

Now try this

Diagram **NOT** accurately drawn

4 cm 5 cm
7 cm
3 cm

(a) Calculate the volume of the triangular prism.
(4 marks)

(b) Show that the total surface area of the prism is 96 cm².
(3 marks)

edexcel

When you are asked to show something then you need to set out your work very neatly, explaining where all your calculations come from.

C
D
E
F
G

Cylinders

To find the SURFACE AREA of a CYLINDER you need to add up the areas of the faces. A cylinder has two flat circular faces and one curved face. When you flatten out the curved face it is shaped like a rectangle.

Surface area = 2 × area of circle + area of rectangle

$$= 2 \times \pi r^2 + 2\pi r \times h$$
$$= 2\pi r^2 + 2\pi r h$$

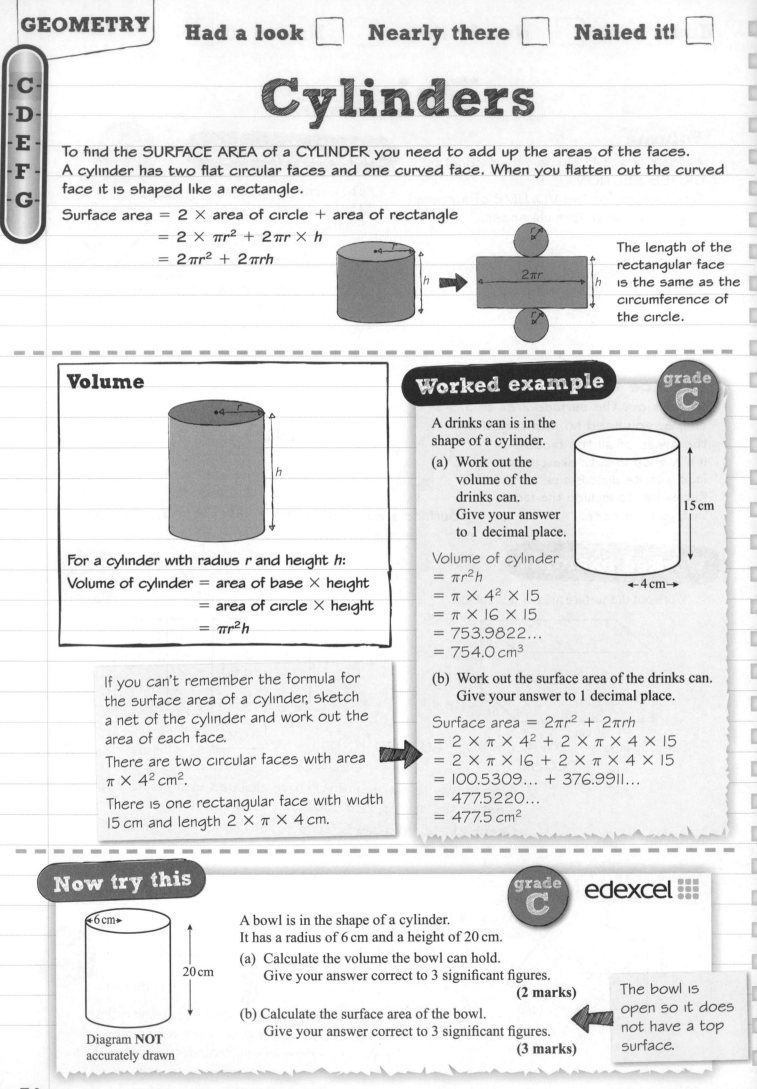

The length of the rectangular face is the same as the circumference of the circle.

Volume

For a cylinder with radius r and height h:

Volume of cylinder = area of base × height

                 = area of circle × height

                 = $\pi r^2 h$

If you can't remember the formula for the surface area of a cylinder, sketch a net of the cylinder and work out the area of each face.

There are two circular faces with area $\pi \times 4^2$ cm².

There is one rectangular face with width 15 cm and length 2 × π × 4 cm.

Worked example

grade **C**

A drinks can is in the shape of a cylinder.

(a) Work out the volume of the drinks can. Give your answer to 1 decimal place.

15 cm

←4 cm→

Volume of cylinder
$= \pi r^2 h$
$= \pi \times 4^2 \times 15$
$= \pi \times 16 \times 15$
$= 753.9822...$
$= 754.0 \text{ cm}^3$

(b) Work out the surface area of the drinks can. Give your answer to 1 decimal place.

Surface area $= 2\pi r^2 + 2\pi r h$
$= 2 \times \pi \times 4^2 + 2 \times \pi \times 4 \times 15$
$= 2 \times \pi \times 16 + 2 \times \pi \times 4 \times 15$
$= 100.5309... + 376.9911...$
$= 477.5220...$
$= 477.5 \text{ cm}^2$

Now try this

grade **C**   edexcel ▦

←6 cm→

20 cm

Diagram **NOT** accurately drawn

A bowl is in the shape of a cylinder. It has a radius of 6 cm and a height of 20 cm.

(a) Calculate the volume the bowl can hold. Give your answer correct to 3 significant figures.

(2 marks)

(b) Calculate the surface area of the bowl. Give your answer correct to 3 significant figures.

(3 marks)

The bowl is open so it does not have a top surface.

74

C
D
E
F
G

Units of area and volume

Converting units of area or volume is trickier than converting units of length. You need to remember your area and volume conversions for your exam.

These two squares have the same area.

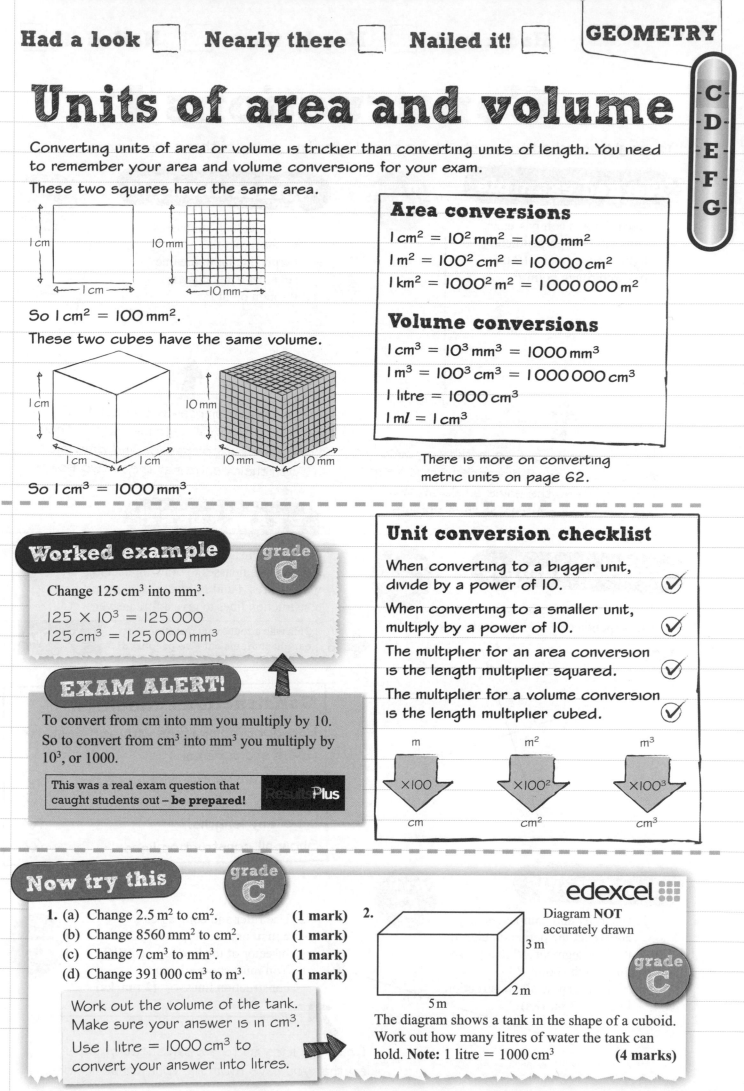

So $1 \, cm^2 = 100 \, mm^2$.

These two cubes have the same volume.

So $1 \, cm^3 = 1000 \, mm^3$.

Area conversions

$1 \, cm^2 = 10^2 \, mm^2 = 100 \, mm^2$

$1 \, m^2 = 100^2 \, cm^2 = 10\,000 \, cm^2$

$1 \, km^2 = 1000^2 \, m^2 = 1\,000\,000 \, m^2$

Volume conversions

$1 \, cm^3 = 10^3 \, mm^3 = 1000 \, mm^3$

$1 \, m^3 = 100^3 \, cm^3 = 1\,000\,000 \, cm^3$

$1 \, litre = 1000 \, cm^3$

$1 \, ml = 1 \, cm^3$

There is more on converting metric units on page 62.

Worked example

grade **C**

Change $125 \, cm^3$ into mm^3.

$125 \times 10^3 = 125\,000$

$125 \, cm^3 = 125\,000 \, mm^3$

EXAM ALERT!

To convert from cm into mm you multiply by 10. So to convert from cm^3 into mm^3 you multiply by 10^3, or 1000.

This was a real exam question that caught students out – **be prepared!**

ResultsPlus

Unit conversion checklist

When converting to a bigger unit, divide by a power of 10. ✓

When converting to a smaller unit, multiply by a power of 10. ✓

The multiplier for an area conversion is the length multiplier squared. ✓

The multiplier for a volume conversion is the length multiplier cubed. ✓

m m^2 m^3

×100 $×100^2$ $×100^3$

cm cm^2 cm^3

Now try this

grade **C**

edexcel ▦

1. (a) Change $2.5 \, m^2$ to cm^2.   **(1 mark)**

   (b) Change $8560 \, mm^2$ to cm^2.   **(1 mark)**

   (c) Change $7 \, cm^3$ to mm^3.   **(1 mark)**

   (d) Change $391\,000 \, cm^3$ to m^3.   **(1 mark)**

Work out the volume of the tank. Make sure your answer is in cm^3. Use $1 \, litre = 1000 \, cm^3$ to convert your answer into litres.

2.

Diagram **NOT** accurately drawn

3 m

2 m

5 m

grade **C**

The diagram shows a tank in the shape of a cuboid. Work out how many litres of water the tank can hold. **Note:** 1 litre = $1000 \, cm^3$ **(4 marks)**

C
D
E
F
G

Constructions 1

You might be asked to construct a perpendicular line in any of these three ways.

Worked example
grade C

Use ruler and compasses to **construct** the perpendicular to the line segment AB that passes through point P.

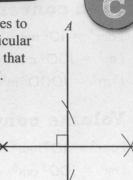

Use your compasses to mark two points on the line an equal distance from P. Keep the compasses the same and draw two arcs with their centres at these points.

Worked example
grade C

Use ruler and compasses to **construct** the perpendicular to the line segment AB that passes through point P.

Use your compasses to mark two points an equal distance from P. Then widen your compasses and draw arcs with their centres at these two points.

Worked example
grade C

Use ruler and compasses to **construct** the perpendicular bisector of the line AB.

EXAM ALERT!

Only 15% of students got full marks on this question. Use your compasses to draw intersecting arcs with centres A and B. You must show **all** your construction lines to give a full answer.

This was a real exam question that caught students out – **be prepared!**   ResultsPlus

Constructions checklist

Use good compasses with stiff arms. ✓
Use a sharp pencil. ✓
Use a transparent ruler. ✓
Mark any angles. ✓
Label any lengths. ✓
Show all construction lines. ✓

Now try this
grade C

edexcel

1. Use ruler and compasses to **construct** the perpendicular to the line segment AB that passes through the point P.
You must show all construction lines. **(2 marks)**

2. Use ruler and compasses to **construct** the perpendicular bisector of the line segment PQ.
You must show all your construction lines. **(2 marks)**

P ——————————— Q

Copy the lines on to a clean sheet of paper so that you have room for your constructions.

Constructions 2

You need to know all of these constructions for your exam.

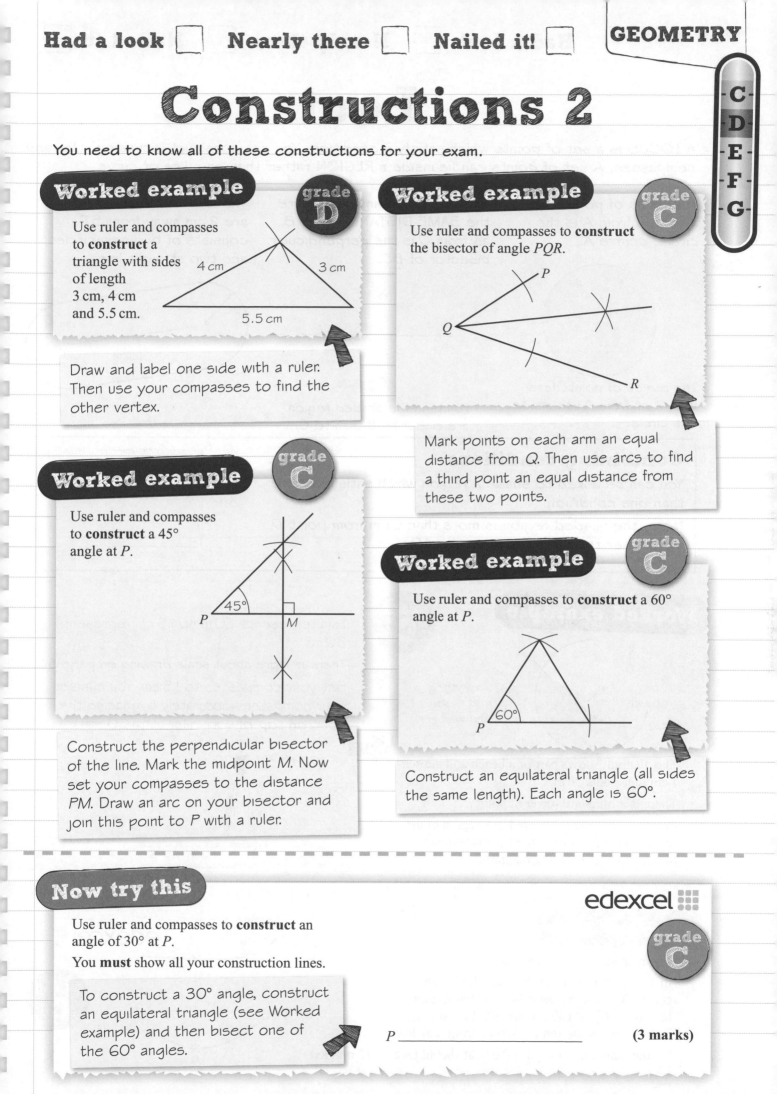

C
D
E
F
G

Worked example

grade **D**

Use ruler and compasses to **construct** a triangle with sides of length 3 cm, 4 cm and 5.5 cm.

4 cm   3 cm

5.5 cm

Draw and label one side with a ruler. Then use your compasses to find the other vertex.

Worked example

grade **C**

Use ruler and compasses to **construct** the bisector of angle PQR.

P

Q

R

Mark points on each arm an equal distance from Q. Then use arcs to find a third point an equal distance from these two points.

Worked example

grade **C**

Use ruler and compasses to **construct** a 45° angle at P.

45°

P   M

Construct the perpendicular bisector of the line. Mark the midpoint M. Now set your compasses to the distance PM. Draw an arc on your bisector and join this point to P with a ruler.

Worked example

grade **C**

Use ruler and compasses to **construct** a 60° angle at P.

60°

P

Construct an equilateral triangle (all sides the same length). Each angle is 60°.

Now try this

edexcel ⦂⦂⦂

Use ruler and compasses to **construct** an angle of 30° at P.

You **must** show all your construction lines.

grade **C**

To construct a 30° angle, construct an equilateral triangle (see Worked example) and then bisect one of the 60° angles.

P _____

(3 marks)

Loci

A LOCUS is a set of points which satisfy a condition. You can construct loci using ruler and compasses. A set of points can lie inside a REGION rather than on a line or curve.

The locus of points which are 7 cm from A is the circle, centre A.

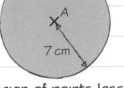

The region of points less than 7 cm from A lies inside this circle.

The locus of points which are the SAME DISTANCE from B as from C is the perpendicular bisector of BC.

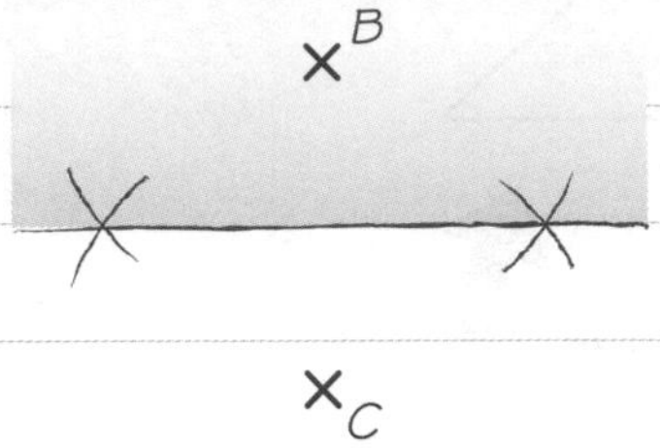

Points in the shaded region are closer to B than to C.

The locus of points which are 2 cm away from ST consists of two semicircles and two straight lines.

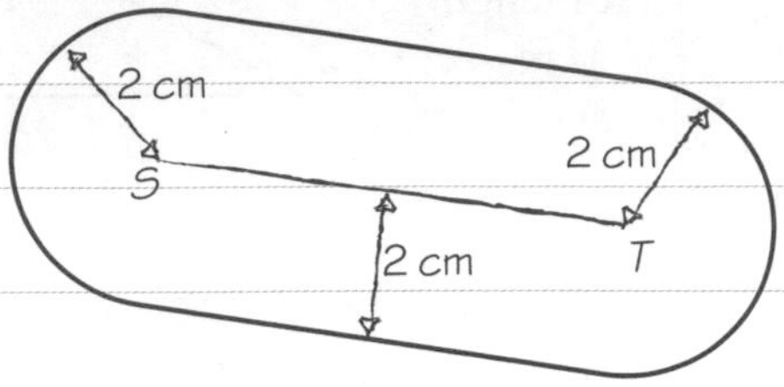

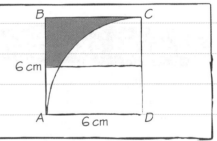

Combining conditions

You can be asked to shade a region which satisfies more than one condition.

Here, the shaded region is more than 6 cm from point D and closer to line BC than to line AD.

Worked example   grade **C**

Beach   P×

Sea

Everything in red is part of the answer.

The diagram shows part of a beach and the sea. 1 cm represents 20 m.

There is a lifeguard tower at point P.

Public swimming is allowed in a region of the sea less than 30 m from the lifeguard tower. Shade this region on the diagram.

I cm represents 20 m so 1.5 cm represents 30 m.

There is more about scale drawing on page 56.

Set your compasses to 1.5 cm. You can set your compasses accurately by placing the point **on top of** your ruler at the O mark.

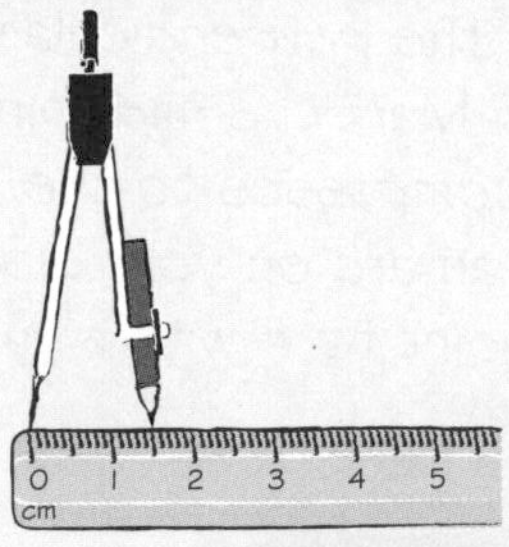

Now try this   edexcel

The map shows part of a lake.

In a competition for radio-controlled boats, a competitor has to steer a boat so that its path between *AB* and *CD* is a straight line **and** this path is always the same distance from *A* as from *B*.

On the map, draw the path the boat should take.   **(2 marks)**

grade **C**

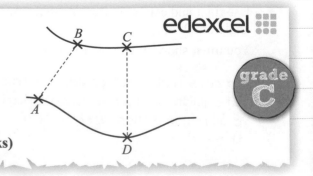

Translations

A translation is a sliding movement. You can describe a translation using a VECTOR.

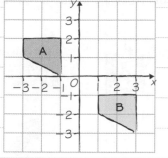

The transformation A → B is a translation by the vector

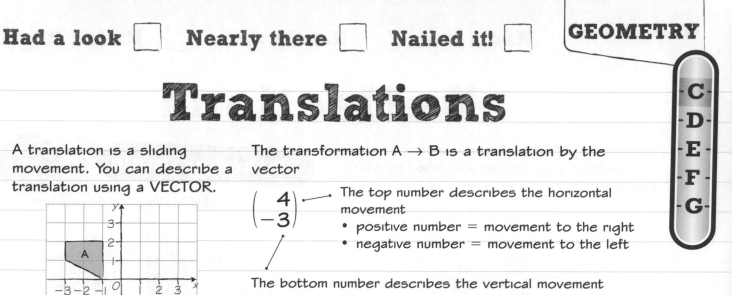

$\begin{pmatrix} 4 \\ -3 \end{pmatrix}$

The top number describes the horizontal movement
- positive number = movement to the right
- negative number = movement to the left

The bottom number describes the vertical movement
- positive number = movement up
- negative number = movement down

Translated shapes are CONGRUENT.
For a reminder about congruent shapes have a look at page 59.

Worked example   grade C

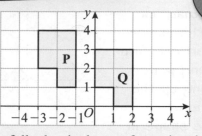

Describe <u>fully</u> the single transformation that will map shape **P** onto shape **Q**.

Translation by the vector $\begin{pmatrix} 3 \\ -1 \end{pmatrix}$

For translations, lengths of sides do not change and angles of shape do not change.

EXAM ALERT!

Hardly any students got both marks for this question.

First write the word 'translation'. Do **not** use words such as 'transformed' or 'moved'.

You must use a vector to describe the translation:
$\begin{pmatrix} movement\ to\ the\ right \\ movement\ up \end{pmatrix}$

Use negative numbers to describe movement to the left or down.

Do **not** describe the movement as 'across 3 and down 1'.

This was a real exam question that caught students out – **be prepared!** ResultsPlus

Now try this   edexcel

1.   grade C

On the grid, translate the shaded shape **P** by the vector $\begin{pmatrix} -2 \\ -3 \end{pmatrix}$.
Label the new shape **R**.   **(2 marks)**

2.

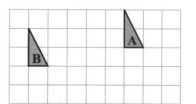

Describe fully the single transformation that will map shape **A** onto shape **B**.   **(2 marks)**

grade C

Read the question carefully to see where you start.

C
D
E
F
G

Reflections

You can REFLECT a shape in a mirror line. To describe a reflection you need to give the EQUATION of the mirror line.

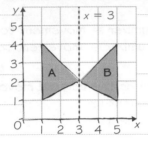

The transformation A → B is a reflection in the line $x = 3$.

Reflected shapes are CONGRUENT.

Worked example

Reflect the shaded shape in the mirror line.

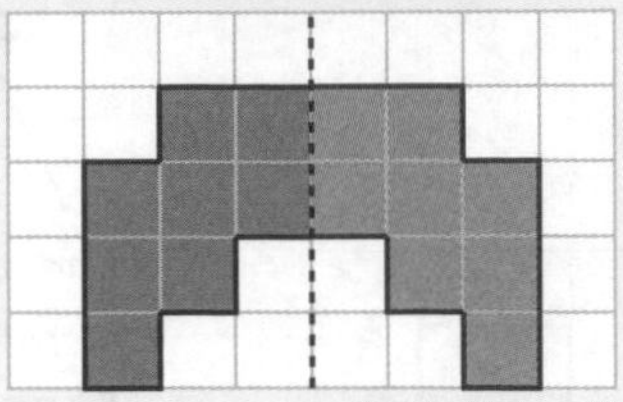

Everything in red is part of the answer.

The mirror line is a line of symmetry.

Quick reflections

It is easy to reflect shapes and check your answers using tracing paper.

Trace the original shape including the mirror line.

Turn the diagram so that the mirror line is vertical. Turn the tracing paper over, lining up the mirror lines.
Trace the shape in the new position.

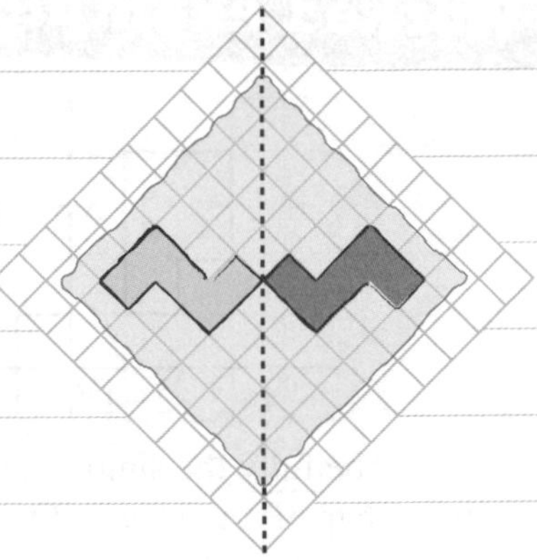

Worked example

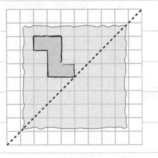

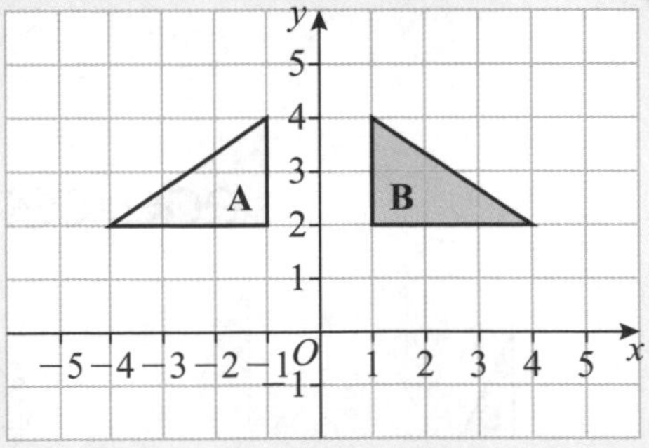

Describe fully the single transformation that will map shape **A** onto shape **B**.

Reflection in the y-axis.

The question says 'describe fully'. You need to write
• the word 'reflection'
• a description of the mirror line.
You can describe the mirror line using an equation, or by writing 'y-axis' or 'x-axis'.

Now try this

edexcel

1. Reflect the shaded shape in the mirror line.   **(2 marks)**

mirror line

2. Describe fully the transformation that maps triangle **A** onto triangle **B**.   **(2 marks)**

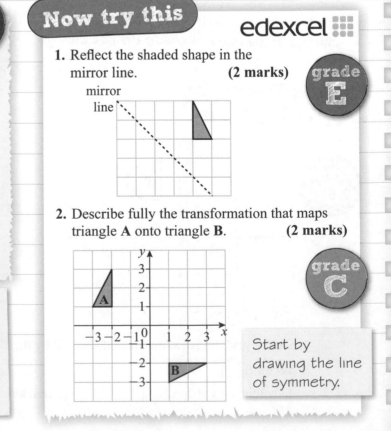

Start by drawing the line of symmetry.

Rotations

To describe a ROTATION you need to give

- the centre of rotation
- the angle of rotation
- the direction of rotation.

The centre of rotation is often the origin O. Otherwise it is given as coordinates.

The angle of rotation is given as 90° (one quarter turn) or 180° (one half turn).

The direction of rotation is given as clockwise or anticlockwise.

WATCH OUT! You don't need to give a direction for a rotation of 180°.

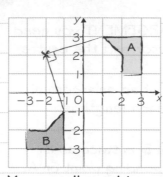

A to B: Rotation 90° clockwise about the point (−2, 2).

You could also say 'a quarter turn clockwise about (−2, 2)'.

You are allowed to ask for tracing paper in the exam. This makes it really easy to rotate shapes and check your answers.

Rotated shapes are CONGRUENT.

For a reminder about congruent shapes have a look at page 59.

Worked example grade C

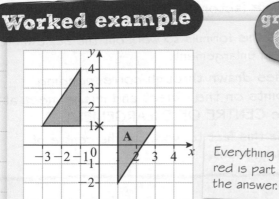

Everything in red is part of the answer.

On the grid, rotate the shaded shape **A** one half turn about (0, 1).

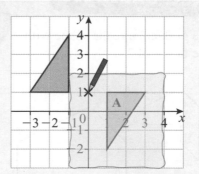

Mark the centre of rotation (0, 1) with a ×.

Trace the shape and put your pencil or compass on the ×. Rotate the tracing paper to rotate the shape.

Now try this

edexcel

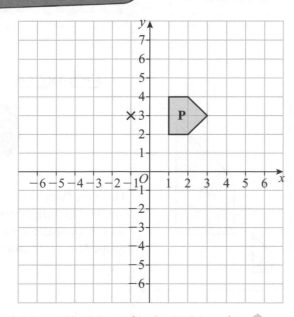

On the grid,

(a) rotate the shaded shape **P** one quarter turn clockwise about O. Label the new shape **Q**.
   (3 marks)

(b) rotate the shaded shape **P** 90° anticlockwise about (−1, 3). Label the new shape **R**.
   (3 marks)

In (b), make sure you use the given centre of rotation.

grade C

C
D
E
F
G

Enlargements

To describe an ENLARGEMENT you need to give the scale factor and the centre of enlargement.

The SCALE FACTOR of an enlargement tells you how much each length is multiplied by.

A to B: Each point on B is twice as far from C as the corresponding point on A.

The transformation A → B is an enlargement with scale factor 2, centre (1, 4).

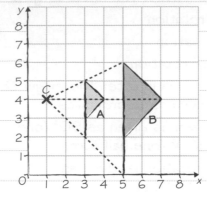

Worked example

grade D

On the grid, enlarge the shape with a scale factor of 2

Describing enlargements

Scale factor = $\dfrac{\text{enlarged length}}{\text{original length}}$

Use this formula to work out the scale factor of an enlargement.

Lines drawn through corresponding points on the object and image meet at the CENTRE OF ENLARGEMENT.

Use this fact to work out the centre of enlargement.

If no centre of enlargement is given, you can draw your new shape anywhere on the grid. Just make sure that every length on the enlarged shape is **2 times** the corresponding length on the original shape.

Now try this

grade D

edexcel

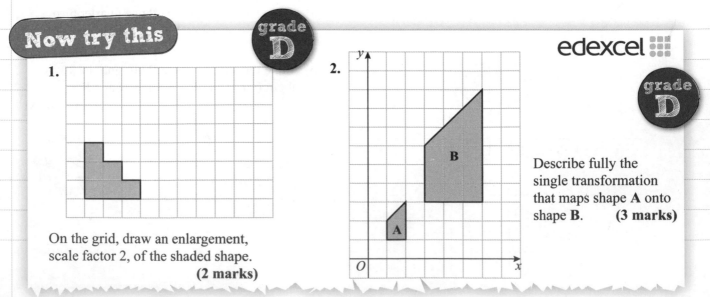

1.

On the grid, draw an enlargement, scale factor 2, of the shaded shape.
(2 marks)

2.

Describe fully the single transformation that maps shape **A** onto shape **B**.   **(3 marks)**

grade D

Combining transformations

C D E F G

You can describe two transformations using a single transformation.

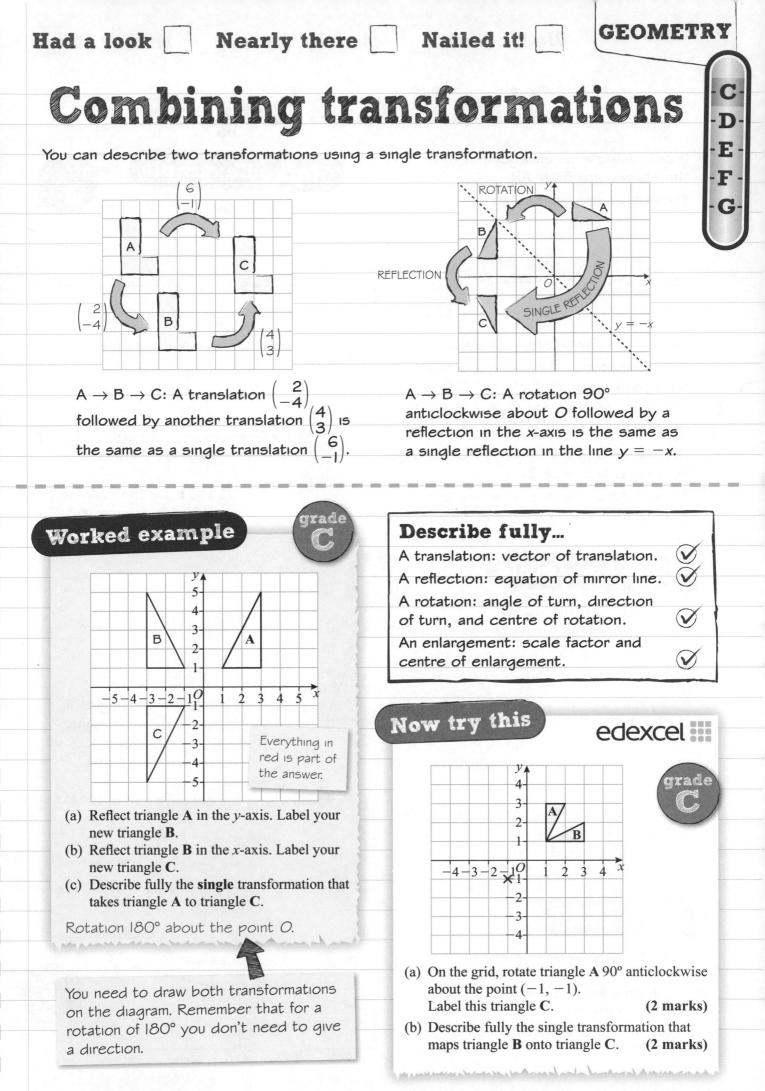

$A \rightarrow B \rightarrow C$: A translation $\begin{pmatrix} 2 \\ -4 \end{pmatrix}$ followed by another translation $\begin{pmatrix} 4 \\ 3 \end{pmatrix}$ is the same as a single translation $\begin{pmatrix} 6 \\ -1 \end{pmatrix}$.

$A \rightarrow B \rightarrow C$: A rotation 90° anticlockwise about O followed by a reflection in the x-axis is the same as a single reflection in the line $y = -x$.

Worked example

grade **C**

Everything in red is part of the answer.

(a) Reflect triangle **A** in the y-axis. Label your new triangle **B**.
(b) Reflect triangle **B** in the x-axis. Label your new triangle **C**.
(c) Describe fully the **single** transformation that takes triangle **A** to triangle **C**.

Rotation 180° about the point O.

You need to draw both transformations on the diagram. Remember that for a rotation of 180° you don't need to give a direction.

Describe fully...

A translation: vector of translation. ✓
A reflection: equation of mirror line. ✓
A rotation: angle of turn, direction of turn, and centre of rotation. ✓
An enlargement: scale factor and centre of enlargement. ✓

Now try this

edexcel

grade **C**

(a) On the grid, rotate triangle **A** 90° anticlockwise about the point $(-1, -1)$. Label this triangle **C**. **(2 marks)**
(b) Describe fully the single transformation that maps triangle **B** onto triangle **C**. **(2 marks)**

Similar shapes

C
D
E
F
G

If one shape is an enlargement of another, the shapes are SIMILAR.

These triangles are similar.

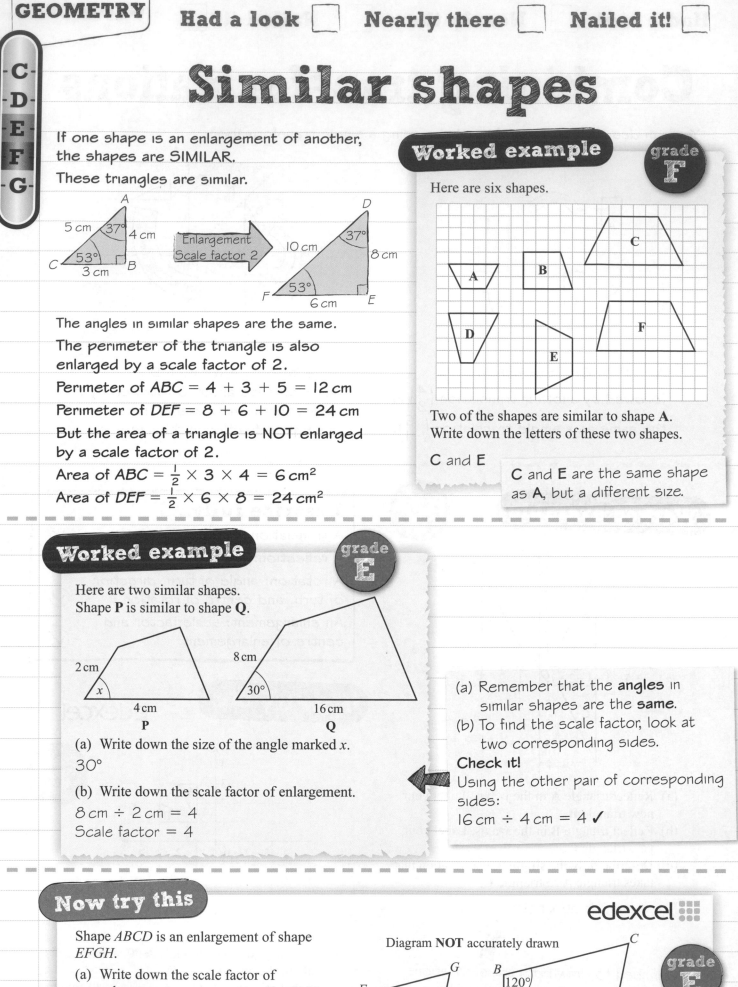

The angles in similar shapes are the same.

The perimeter of the triangle is also enlarged by a scale factor of 2.

Perimeter of ABC = 4 + 3 + 5 = 12 cm

Perimeter of DEF = 8 + 6 + 10 = 24 cm

But the area of a triangle is NOT enlarged by a scale factor of 2.

Area of ABC = $\frac{1}{2} \times 3 \times 4 = 6\,cm^2$

Area of DEF = $\frac{1}{2} \times 6 \times 8 = 24\,cm^2$

Worked example

grade **F**

Here are six shapes.

Two of the shapes are similar to shape **A**. Write down the letters of these two shapes.

C and E

C and E are the same shape as A, but a different size.

Worked example

grade **E**

Here are two similar shapes.
Shape **P** is similar to shape **Q**.

(a) Write down the size of the angle marked x.

30°

(b) Write down the scale factor of enlargement.

8 cm ÷ 2 cm = 4

Scale factor = 4

(a) Remember that the **angles** in similar shapes are the **same**.
(b) To find the scale factor, look at two corresponding sides.

Check it!
Using the other pair of corresponding sides:
16 cm ÷ 4 cm = 4 ✓

Now try this

edexcel

Shape *ABCD* is an enlargement of shape *EFGH*.

(a) Write down the scale factor of enlargement.   **(1 mark)**

(b) Write down the size of the angle marked x.   **(1 mark)**

Diagram **NOT** accurately drawn

grade **E**

C
D
E
F
G

Pythagoras' theorem

Pythagoras' theorem is a really useful rule. You can use it to find the length of a missing side in a right-angled triangle.

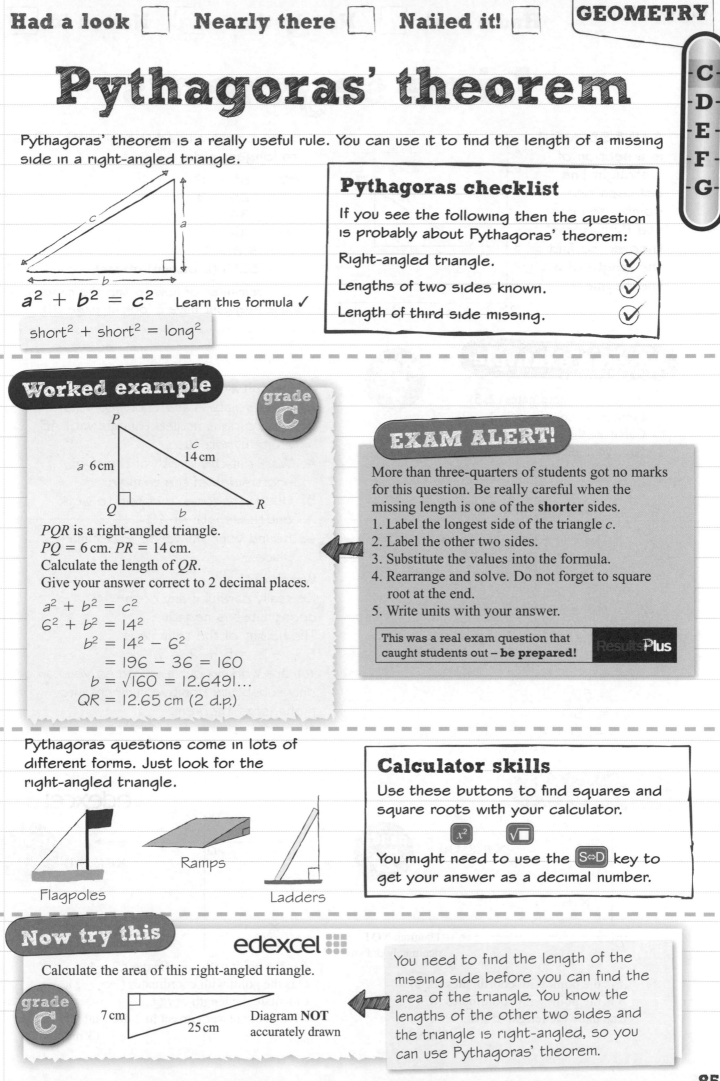

$$a^2 + b^2 = c^2$$ Learn this formula ✓

short² + short² = long²

Pythagoras checklist

If you see the following then the question is probably about Pythagoras' theorem:

Right-angled triangle.   ✓

Lengths of two sides known.   ✓

Length of third side missing.   ✓

Worked example

grade **C**

PQR is a right-angled triangle.
$PQ = 6$ cm. $PR = 14$ cm.
Calculate the length of *QR*.
Give your answer correct to 2 decimal places.

$a^2 + b^2 = c^2$
$6^2 + b^2 = 14^2$
$\quad\quad b^2 = 14^2 - 6^2$
$\quad\quad\quad = 196 - 36 = 160$
$\quad\quad b = \sqrt{160} = 12.6491...$
$\quad QR = 12.65$ cm (2 d.p.)

EXAM ALERT!

More than three-quarters of students got no marks for this question. Be really careful when the missing length is one of the **shorter** sides.
1. Label the longest side of the triangle *c*.
2. Label the other two sides.
3. Substitute the values into the formula.
4. Rearrange and solve. Do not forget to square root at the end.
5. Write units with your answer.

This was a real exam question that caught students out – **be prepared!**   ResultsPlus

Pythagoras questions come in lots of different forms. Just look for the right-angled triangle.

Flagpoles   Ramps   Ladders

Calculator skills

Use these buttons to find squares and square roots with your calculator.

x^2 $\sqrt{\square}$

You might need to use the S⇔D key to get your answer as a decimal number.

Now try this

edexcel

Calculate the area of this right-angled triangle.

grade **C**

7 cm   25 cm   Diagram **NOT** accurately drawn

You need to find the length of the missing side before you can find the area of the triangle. You know the lengths of the other two sides and the triangle is right-angled, so you can use Pythagoras' theorem.

Line segments

A LINE SEGMENT is a section of a straight line between two points. You can use Pythagoras' theorem to find the length of a line segment.

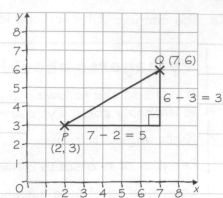

Draw a right-angled triangle with PQ as its longest side.

$$PQ^2 = 5^2 + 3^2$$
$$= 25 + 9$$
$$= 34$$
$$PQ = \sqrt{34}$$
$$= 5.8309...$$
$$= 5.83 \text{ (2 decimal places)}$$

For a reminder of how to find the midpoint of a line segment have a look at page 39.

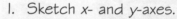

Worked example

grade **C**

Point A has coordinates $(2, 5)$.
Point B has coordinates $(3, -2)$.
Calculate the length of the line segment AB.
Give your answer correct to 2 decimal places.

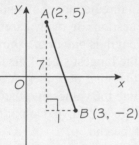

Horizontal distance $= 3 - 2 = 1$
Vertical distance $= 5 - -2 = 7$
$AB^2 = 1^2 + 7^2$
$\quad = 50$
$AB = \sqrt{50}$
$\quad = 7.07106... = 7.07 \text{ (2 d.p.)}$

1. Sketch x- and y-axes.
2. Mark points A and B on your sketch.
3. Draw a right-angled triangle with AB as its hypotenuse.
4. Work out the length of the two short sides of the triangle.
5. Use Pythagoras' theorem to work out the length of AB.
6. Round your answer to 2 decimal places.

Watch out!

Be really careful if any of the coordinates is negative.
The height of the triangle is $5 - -2 = 5 + 2 = 7$.
You are working out a length so you can only substitute positive numbers into Pythagoras' theorem.

Now try this

edexcel

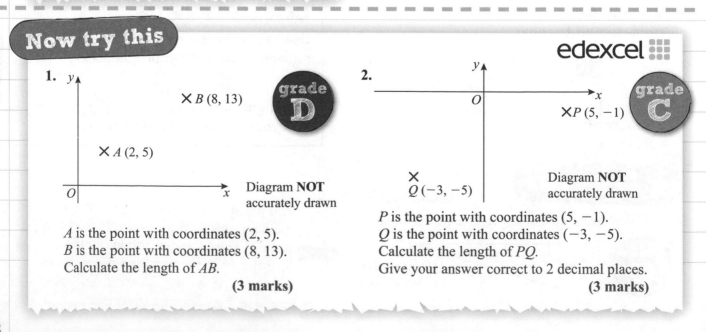

1.
A is the point with coordinates $(2, 5)$.
B is the point with coordinates $(8, 13)$.
Calculate the length of AB.

grade **D**

Diagram **NOT** accurately drawn

(3 marks)

2.
P is the point with coordinates $(5, -1)$.
Q is the point with coordinates $(-3, -5)$.
Calculate the length of PQ.
Give your answer correct to 2 decimal places.

grade **C**

Diagram **NOT** accurately drawn

(3 marks)

Problem-solving practice

About half of the questions on your exam will need problem-solving skills.

These skills are sometimes called AO2 and AO3.

You can use a calculator on question 5, but practise the other questions without one. You might have to answer similar questions on your non-calculator paper.

For these questions you might need to:

- choose what method to use
- use the maths you've learnt in a new context
- plan your answer when solving harder problems
- show your working clearly and give reasons for your answers.

AO2

AO3

---

1 On the grid, show how the shaded shape will tessellate.

(2 marks)

Congruent shapes p. 59

grade **E**

Shapes which **tessellate** cover an area with no gaps or overlaps. You need to draw copies of the shape which are **congruent**, that is, the same shape and size. You can reflect or rotate the shape.

TOP TIP

Most tessellations follow some sort of pattern.

2

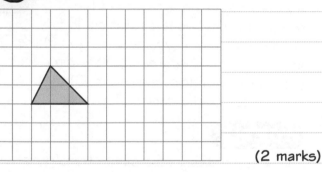

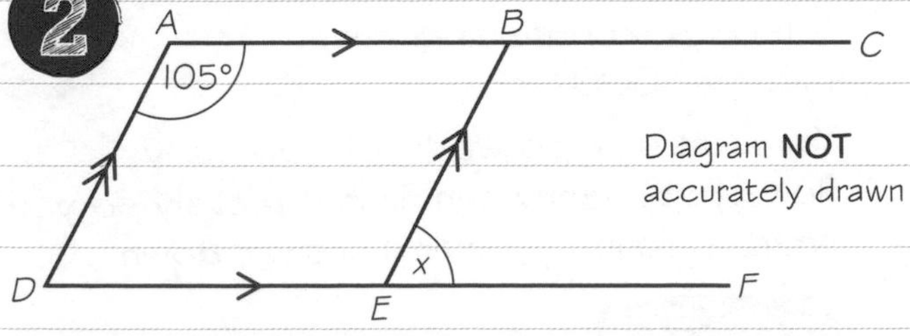

Diagram **NOT** accurately drawn

ABED is a parallelogram.

ABC and *DEF* are straight lines.

Find the size of the angle marked *x*.

You must give reasons to explain your answer.

(4 marks)

Angles 2 p. 51

grade **D**

Just writing down the size of angle *x* is not enough. For **each step** of your working, write down any angle you have worked out **and** the angle fact or property you used.

TOP TIP

It's a good idea to write missing angles on the diagram as you work them out, but you still need to write down the REASON for each step of your working.

Problem-solving practice

3 A buoy is 6 km from a ship on a bearing of 290°.

A lighthouse is 8 km east of the ship.

Work out the distance between the buoy and the lighthouse.   (3 marks)

grade **C**

Bearings p. 55
Scale drawings and maps p. 56

You'll need to draw a scale diagram to solve this question. A good scale to use would be 1 cm = 2 km. Always use a ruler and a sharp pencil to draw any lines and **don't** rub out any construction lines or working.

TOP TIP

Bearings less than 180° are to the RIGHT of North. Bearings between 180° and 360° are to the LEFT. Bearings always have three figures.

4
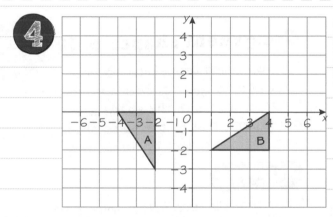

Describe fully the single transformation which maps triangle A onto triangle B.
                    (3 marks)

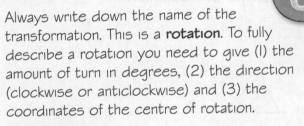
grade **C**

Rotations p. 81

Always write down the name of the transformation. This is a **rotation**. To fully describe a rotation you need to give (1) the amount of turn in degrees, (2) the direction (clockwise or anticlockwise) and (3) the coordinates of the centre of rotation.

TOP TIP

You need to describe a **single** transformation. You can't say that the shape is rotated then moved to the right. Use tracing paper to find a centre of rotation that maps triangle **A** directly onto triangle **B**.

---

5 *
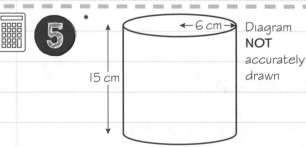
←6 cm→ Diagram NOT accurately drawn

15 cm

Jenny fills some empty flowerpots completely with compost.

Each flowerpot is in the shape of a cylinder of height 15 cm and radius 6 cm.

Jenny has a 15 litre bag of compost.

She fills up each flowerpot completely.

How many flowerpots can she fill completely?

You must show your working.   (4 marks)

grade **C**

Cylinders p. 74

You need to remember that 1 litre = 1000 cm³.

You are trying to work out how many flowerpots Jenny can fill **completely** so you'll need to round your final answer **down**.

TOP TIP

If a question has a * next to it, then there are marks available for QUALITY OF WRITTEN COMMUNICATION. This means you must show all your working and write your answer clearly with the correct units.

Collecting data

You need to know what makes a question good or bad in a survey.
Look at this example then read the comments.

Internet use survey

1. What do you use to access the internet?
   ....

✗ It's hard to know what this question means. Add at least 4 response boxes to improve the question:
☐ Laptop computer
☐ Desktop computer
☐ Smartphone
☐ Other

2. How much time do you spend on the internet each day?
   ☐ Not very much   ☐ Average   ☐ A lot

✗ These responses could mean different things to different people. It would be better to ask how many hours they spend on the internet each day.

3. How many times a week do you check your email?
   ☐ 1–5   ☐ 5–10   ☐ 10–15   ☐ Every day

✗ The response boxes overlap. These response boxes would be better:
☐ 0–5   ☐ 6–10
☐ 11–15
☐ 16 or more

4. Have you ever downloaded films illegally from the internet?
   ....

✗ People aren't very likely to answer this question truthfully. Don't ask people to reveal embarrassing or personal information.

5. Do you agree that the BBC iPlayer is very easy to use?
   ☐ Yes   ☐ No

✗ This is a **biased** question. People are more likely to agree with you. A better question would be: The BBC iPlayer is easy to use.
☐ Agree   ☐ Disagree
☐ Neither

Worked example   grade D

Katya wants to find out information about the numbers of men, women, boys and girls using a sports hall.

Design a suitable <u>data collection sheet</u> to collect the information.

	Tally	Frequency
Men		
Women		
Boys		
Girls		

Everything in red is part of the answer.

EXAM ALERT!

Only 40% of students got full marks for this question. Make sure you design a **data collection sheet** (**tally chart**) and **not** a questionnaire.
Your data collection sheet must include columns for 'Tally' and 'Frequency'.

This was a real exam question that caught students out – **be prepared!**   Results Plus

Now try this   edexcel

1. Toby wants to find out how many text messages people send.
   He uses this question on a questionnaire.

   | How many text messages have you sent on your mobile? |
   | 0–10   10–20   20–30   30 or more |

   (a) Write down **two** things wrong with this question.   **(2 marks)**   grade D

   Toby also wants to find out how much time people spend talking on their mobile phones.

   (b) Design a suitable question Toby could use for his questionnaire. You must include some response boxes.   **(2 marks)**   grade C

In part (b), make sure that your response boxes cover every possible answer.

C
D
E
F
G

Two-way tables

A TWO-WAY TABLE is a special type of frequency table which has a heading for each row and each column.

	Year 7	Year 8	Year 9	Total
Vegetarian	14	22	25	61
Not vegetarian	72	63	54	189
Total	86	85	79	250

There were 61 vegetarians in total.

In total 250 students were surveyed.

There were 86 Year 7 students surveyed.

There were 63 non-vegetarians in Year 8.

Worked example

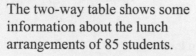

The two-way table shows some information about the lunch arrangements of 85 students.

	School lunch	Packed lunch	Other	Total
Female	21	13	13	47
Male	19	5	14	38
Total	40	18	27	85

Complete the two-way table.

'School lunch' column: 40 − 21 = 19
'Female' row: 47 − 21 − 13 = 13
'Packed lunch' column: 13 + 5 = 18
'Total' row: 85 − 40 − 18 = 27
'Other' column: 27 − 13 = 14
'Male' row: 19 + 5 + 14 = 38

Check:
47 + 38 = 85
40 + 18 + 27 = 85

Golden rules

The numbers in each column add up to the total for that column.

Other
13
+ 14
= 27

The numbers in each row add up to the total for that row.

Female	21	+ 13	+ 13	= 47

You might have to complete a two-way table in the exam.

1. Look for rows or columns with only one missing number.
2. Use subtraction to find any missing numbers in the table.
3. Use addition to find any missing totals.
4. Fill in the missing values as you go along.

Check it!
Add up the row totals and the column totals. They should be the same.

Now try this

Draw a two-way table for this information. Fill in all the given numbers and then complete the table.

edexcel

grade D

1. The two-way table shows some information about the numbers of students in a school.

	Year group			Total
	9	**10**	**11**	
Boys			125	407
Girls		123		
Total	303	256		831

Complete the two-way table.   **(3 marks)**

grade C

2. 80 children went on a school trip. They went to London or to York. 23 boys and 19 girls went to London. 14 boys went to York.

(a) Draw a table or chart to show this information.

(b) How many girls went to York?   **(3 marks)**

Pictograms

A PICTOGRAM can be used to represent data from a tally chart or frequency table.
This pictogram shows the results of a survey about how people watch television.
There is one row for each option.

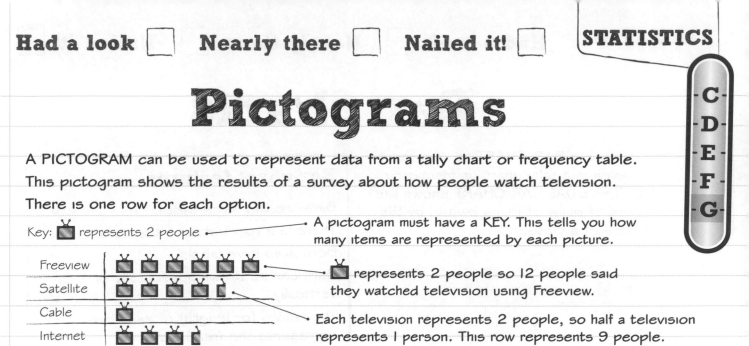

Key: 📺 represents 2 people •————→ A pictogram must have a KEY. This tells you how many items are represented by each picture.

Freeview	📺 📺 📺 📺 📺 📺
Satellite	📺 📺 📺 📺 📺
Cable	📺
Internet	📺 📺 📺 📺

📺 represents 2 people so 12 people said they watched television using Freeview.

Each television represents 2 people, so half a television represents 1 person. This row represents 9 people.

To work out the total number of people in the survey, add together the totals of each row: 12 + 9 + 2 + 7 = 30

Worked example

grade G

The pictogram shows the numbers of films Adam watched in June, July and August.

June	⊞ ⊞
July	⊞ ⊟
August	⊞ ▫
September	⊞ ⊞ ⊟

Everything in red is part of the answer.

Key: ⊞ represents 4 films

(a) Write down the number of films Adam watched in June.

8

(b) Write down the number of films Adam watched in August.

5

In September Adam watched 10 films.
(c) Use this information to complete the pictogram.

Use the key to work out what each picture represents.

⊞ = 4 films   ⊟ = 3 films
⊟ = 2 films   ▫ = 1 film

There is a block of 4 squares and a block of 1 square in August. This represents 4 + 1 = 5 films.

To represent 10 films you need two blocks of 4 squares and one block of 2 squares. Draw the squares neatly on the pictogram.

Now try this

edexcel ▦

The pictogram gives information about the number of goals scored in a local football league in each of 3 weeks.

First week	⚽ ⚽ ⚽
Second week	⚽ ⚽ ⚽ ⚽
Third week	⚽ ◖
Fourth week	
Fifth week	

Key: ⚽ represents 4 goals

(a) Find the number of goals scored in the first week. **(1 mark)**
(b) Find the number of goals scored in the third week. **(1 mark)**

8 goals were scored in the fourth week.
5 goals were scored in the fifth week.
(c) Show this information on the pictogram. **(2 marks)**

grade G

C
D
E
F
G

Bar charts

You can use a BAR CHART to represent data given in a tally chart or frequency table. This DUAL BAR CHART shows the numbers of pairs of jeans owned by the members of a class.

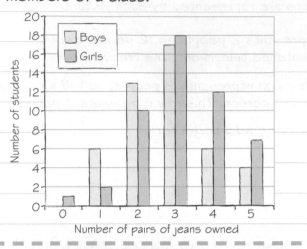

Bar chart features

Bars are the same width.	✓
There is a gap between the bars.	✓
Both axes have labels.	✓
Bars can be drawn horizontally or vertically.	✓
The height (or length) of each bar represents the frequency.	✓
In a dual bar chart two (or more) bars are drawn side by side. They can be used to compare data.	✓

Worked example

grade
F

Kaitlyn carried out a survey of the colours of cars which passed the school gate in 10 minutes. Here are her results.

Colour	Tally
Red	卌 ||||
Black	卌 ||
Silver	卌 卌
Blue	|||

Use the grid to draw a <u>suitable chart or diagram</u> to represent Kaitlyn's results.

The question says 'suitable chart or diagram'. You could also represent this data using a pictogram or pie chart. To find the total number of cars from the bar chart, add up the values of each bar: 3 + 10 + 7 + 9 = 29.

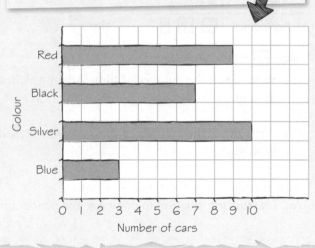

Now try this

grade
G

edexcel

The bar chart shows information about the amounts of time that Andrew and Karen spent watching television on four days last week.

(a) How many minutes did Karen spend watching television on Friday? **(1 mark)**

Karen spent more time watching television than Andrew on two of these four days.

(b) Write down these two days. **(2 marks)**

(c) Work out the total amount of time Andrew spent watching television on these four days. **(2 marks)**

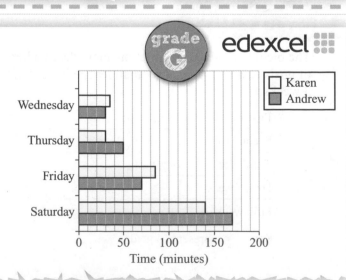

Frequency polygons

C-D-E-F-G

You can represent grouped data using a FREQUENCY POLYGON. Each point is plotted at the MIDPOINT of the class interval. You join the points with straight lines.

Reaction time (r milliseconds)	Frequency
100 ≤ r < 200	7
200 ≤ r < 300	15
300 ≤ r < 400	10

You always record FREQUENCY on the vertical axis.

A reaction time of 146 milliseconds would be in the CLASS INTERVAL 100 ≤ r < 200.

Worked example

30 students ran a cross-country race.
Each student's time was recorded.
The grouped frequency table gives information about these times.

Time (t minutes)	Frequency	Midpoint
10 ≤ t < 14	2	12
14 ≤ t < 18	5	16
18 ≤ t < 22	12	20
22 ≤ t < 26	8	24
26 ≤ t < 30	3	28

Draw a frequency polygon to show this information.

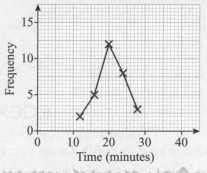

Start by working out the **midpoint** of each class interval. You can add an extra column to your table for this. The midpoint is halfway along the class interval.

The midpoint of the class interval 10 ≤ t < 14 is 12.

Check that your midpoints are all the same distance apart.

Plot the points at the midpoint of each interval. The points are: (12, 2), (16, 5), (20, 12), (24, 8) and (28, 3).

Read the scale on the graph carefully. There are 10 subdivisions between 0 and 5 on the vertical scale, so two small squares represent 1 student.

Join the points with **straight lines**. Do **not** join the two endpoints to each other.

Now try this

grade C

edexcel

60 students take a geography test.
The test is marked out of 50

This table gives information about the students' marks.

Geography mark	0–10	11–20	21–30	31–40	41–50
Frequency	5	11	19	16	9

On the grid, draw a frequency polygon to show this information. **(2 marks)**

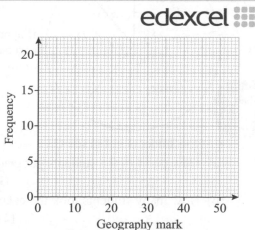

Pie charts

In your exam you might have to draw a PIE CHART from a frequency table, or interpret information given in a pie chart.

There is more about measuring and drawing angles on page 49.

Worked example

grade E

A farm has 40 fruit trees.
The table shows the number of each type of tree.
Draw a pie chart to represent this information.

Type of fruit tree	Number of trees	Angle
Apple	12	12 × 9° = 108°
Plum	5	5 × 9° = 45°
Pear	14	14 × 9° = 126°
Peach	9	9 × 9° = 81°

Angle for 1 tree = 360° ÷ 40 = 9°
Check: 108° + 45° + 126° + 81° = 360° ✓

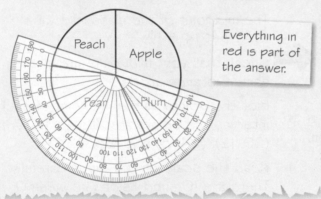

Everything in red is part of the answer.

You need a sharp pencil, compasses and a protractor to draw a pie chart.

1. Add an 'Angle' column to the frequency table.

2. There are 360° in a full circle. There are 40 trees. So divide 360° by 40 to find the angle that represents 1 tree.

3. Multiply the angle that represents 1 tree by the number of each tree type to find the angle for each type.

4. **Check** that your angles add up to 360°.

5. Draw a circle using compasses. Draw a vertical line from the centre to the edge of the circle. Use a protractor to measure and draw the first angle (108°) from this line. Draw each angle carefully in order.

6. Label each sector of your pie chart with the type of fruit tree.

Now try this

edexcel ▦

1. The pie chart gives information about the mathematics exam grades of some students.

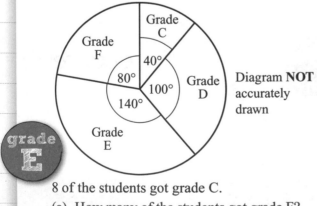

Diagram **NOT** accurately drawn

grade E

8 of the students got grade C.

(a) How many of the students got grade F?
(1 mark)

(b) How many students took the exam?
(2 marks)

2. Ali asked 120 students at his school, 'What is your favourite flavour of crisp?' The table shows his results.

grade D

Flavour of crisp	Frequency
Plain	15
Cheese & Onion	40
Salt & Vinegar	55
Beef	10

Draw an accurate pie chart to show these results. **(4 marks)**

Start off by adding an 'Angle' column to the table. Remember that you'll need a sharp pencil, compasses and a protractor to draw your pie chart.

Averages and range

C
D
E
F
G

There are three different types of average: the mean, median and mode.
The range of a set of data tells you how spread out the data is.

Worked example grade F

The **mode** is the value which occurs **most often**.

The **median** is the middle value. First write the values in order from smallest to largest. If there are two middle values, the median is halfway between them.

Here are six numbers.
4 5 9 7 4 4
(a) Write down the mode.
The mode is 4.
(b) Work out the mean.
4 + 5 + 9 + 7 + 4 + 4 = 33
33 ÷ 6 = 5.5
The mean is 5.5
(c) Work out the median.
4 4 ④ 5 7 9
The median is 4.5
(d) Work out the range.
9 – 4 = 5
The range is 5.

To find the **mean** you add together all the numbers and then divide by how many numbers there are. Don't round your answer.

Range = largest value – smallest value

Worked example grade D

James has three cards.
Each card has a number on it.
The numbers are hidden. ? ? ?
The mode of the three numbers is 4
The mean of the three numbers is 5
Work out the three numbers on the cards.

Mode = 4
At least two cards are 4s.

Mean = 5
Sum of cards = 5 × 3 = 15
4 + 4 + ? = 15
The other card is a 7.
The three cards are 4, 4 and 7.

EXAM ALERT!

Only one-third of students got full marks on this question.

The mode is the most common value.
This means that at least two of the cards must have '4' written on them.   4 4 ?

To find the other number use:
Sum of values = mean × number of values

Check it!
4 + 4 + 7 = 15   4 4 7
15 ÷ 3 = 5 ✓

This was a real exam question that caught students out – **be prepared!**   ResultsPlus

Now try this

For part (d), first work out the range of the girls and the boys.
Then write a sentence to explain why Jenna is correct or incorrect.

edexcel

Here are the test marks of 6 girls and 4 boys.
Girls: 5   3   10   4   7   3
Boys: 2   5   9   3
(a) Write down the mode of the 10 marks.
(1 mark)
(b) Work out the median mark of the boys.
(1 mark)
(c) Work out the mean mark of all 10 students. **(2 marks)**
(d) Jenna says, 'The range of the girls is bigger than the range of the boys.'
Is Jenna correct? Explain why. **(2 marks)**

grade F

C
D
E
F
G

Stem and leaf diagrams

When data is given in a STEM AND LEAF DIAGRAM it is arranged in order of size.

This stem and leaf diagram shows the costs, in £, of some DVDs.

In this diagram, the numbers in the stem represent the number of 'tens'.

Stem	Leaf
0	7 9 9
1	0 0 ② 3 5 7
2	0 5

Key: 1 | 5 = £15

0 | 7 represents £7. This is the smallest data value.

There are 11 pieces of data, so the median is the 6th value. The median is £12.

There are 11 pieces of data in this stem and leaf diagram.

The range of prices is £25 – £7 = £18.

Worked example

grade D

Alison recorded the heights, in cm, of some tree saplings.

35 26 21 23 33 25 42 36
41 26 31 32 40 40 23

Show this information in an ordered stem and leaf diagram.

2	6 1 3 5 6 3
3	5 3 6 1 2
4	2 1 0 0

2	1 3 3 5 6 6
3	1 2 3 5 6
4	0 0 1 2

Key: 2 | 1 represents 21 cm

To draw a stem and leaf diagram:
1. Choose sensible values to use as your stem.
2. Draw an ordered stem, then add the leaves in any order.
3. Cross each data value off the list as you enter it.
4. Redraw the diagram, putting the leaves in order.
5. Add a key.

Worked example

grade E

Eli records the ages of 15 people.
She shows her results in a stem and leaf diagram.

Key: 4 | 3 represents 43 years

1	1 3
2	2 5 5 5
3	1 4 7 7 8 9
4	3 7 9

Eli says, 'The range of the ages is 8 because 9 – 1 = 8'

Eli is **wrong**. Explain why.

49 – 11 = 38
The range is actually 38 years.

EXAM ALERT!

Only half of students got this question right. The easiest way to explain why Eli is wrong is to calculate the correct range.

Look carefully at the key to work out the largest and smallest data values. Then calculate the range. Remember to include the unit (years) in your answer.

This was a real exam question that caught students out – **be prepared!** ResultsPlus

Now try this

edexcel

Jan measures the heights, in mm, of 20 plants in her greenhouse. Here are her results.

178	189	147	147	166
167	153	171	164	158
189	166	165	155	152
147	158	148	151	172

Draw an ordered stem and leaf diagram to show this information. **(3 marks)**

Make the stems 14, 15, 16, 17 and 18.

grade D

Averages from tables 1

C
D
E
F
G

You need to be really careful when you are calculating averages from data given in a frequency table. Questions like this come up almost every year, so make sure you know the method.

Worked example

grade D

Leah asked 40 people how many pets they owned. The table shows her results.

Number of pets x	Frequency f	Frequency × number of pets $f \times x$
0	13	$13 \times 0 = 0$
1	18	$18 \times 1 = 18$
2	7	$7 \times 2 = 14$
3	2	$2 \times 3 = 6$

(a) Write down the mode.

The mode is 1 pet.

(b) Write down the range.

The range is $3 - 0 = 3$ pets.

> Everything in red is part of the answer.

(c) Work out the median.

The median is 1 pet.

(d) Work out the mean.

Total of $f \times x$ column $= 0 + 18 + 14 + 6 = 38$
Total frequency $= 13 + 18 + 7 + 2 = 40$
$38 \div 40 = 0.95$
The mean is 0.95 pets.

(a) The **mode** is the value with the highest frequency. The highest frequency is 18.

(b) The **range** is the difference between the highest and lowest values. So $3 - 0 = 3$ pets.

(c) The **median** is the middle value. There are 40 values so the median is halfway between the 20th and 21st values. The first 13 values are all 0 and the next 18 values are all 1. This means that the 20th and 21st values are both 1, so the median is 1 pet.

(d) To calculate the **mean** from a frequency table you need to add an extra column. Label your new column 'Frequency × number of pets' or '$f \times x$'. The total in the $f \times x$ column represents the total number of pets owned (38 pets).
Use this rule to work out the mean:
$$\text{Mean} = \frac{\text{total of } (f \times x \text{ column})}{\text{total frequency}}$$
Do not round your answer.

Now try this

edexcel

1. Ali found out the number of rooms in each of 40 houses in a town. He used the information to complete the frequency table.

grade E

Number of rooms	Frequency
5	7
6	10
7	12
8	5
9	2

Ali said that the mode is 9 rooms.

(a) Ali is wrong. Explain why. **(1 mark)**

(b) Calculate the total number of rooms. **(2 marks)**

2. Zach has 10 CDs.
The table gives some information about the number of tracks on each CD.

grade D

Number of tracks	Frequency
12	5
13	0
14	2
15	3

(a) Write down the mode. **(1 mark)**

(b) Write down the range. **(1 mark)**

(c) Work out the mean. **(3 marks)**

C
D
E
F
G

Averages from tables 2

Sometimes data in a frequency table is grouped into CLASS INTERVALS. You don't know the exact data values, but you can calculate an ESTIMATE of the mean, and write down which class interval contains the median and which one has the highest frequency.

Worked example

grade **C**

Sethina recorded the times, in minutes, taken to repair 80 car tyres.
Information about these times is shown in the table.

Time (t minutes)	Frequency f	Midpoint x	f × x
$0 < t \leqslant 6$	15	3	$15 \times 3 = 45$
$6 < t \leqslant 12$	24	9	$24 \times 9 = 216$
$12 < t \leqslant 18$	21	15	$21 \times 15 = 315$
$18 < t \leqslant 24$	12	21	$12 \times 21 = 252$
$24 < t \leqslant 30$	8	27	$8 \times 27 = 216$

(a) Write down the modal class interval.

$6 < t \leqslant 12$

(b) Write down which class interval contains the median.

$12 < t \leqslant 18$

(c) Calculate an estimate for the mean time taken to repair each car tyre.

Sum of $f \times x$ column
$= 45 + 216 + 315 + 252 + 216$
$= 1044$
Total frequency
$= 15 + 24 + 21 + 12 + 8 = 80$
$1044 \div 80 = 13.05$ minutes

(a) $0 < t \leqslant 6$ is called a class interval. The class interval with the highest frequency is called the **modal class interval**.

(b) There are 80 data values. So the median is halfway between the 40th and 41st values. There are 39 (15 + 24) values where $t \leqslant 12$ and 60 (15 + 24 + 21) values where $t \leqslant 18$. The median falls in the class interval $12 < t \leqslant 18$.

EXAM ALERT!

90% of students got zero marks when asked to **estimate a mean** from a **grouped frequency table** very similar to this one.

1. Use the first extra column for 'Midpoint (x)'. Work out the midpoint of each class interval.

2. Use the final column for 'f × x'. Multiply each frequency by the midpoint to find the $f \times x$ value. Do **not** use the endpoints of the intervals when calculating $f \times x$.

3. Use this rule to find an estimate for the mean:

$$\text{Estimate of mean} = \frac{\text{total of } (f \times x \text{ column})}{\text{total frequency}}$$

Now try this

First add two extra columns to the table.

The table shows some information about the areas of 50 gardens.

Area of garden (A m²)	Number of gardens (f)
$0 < A \leqslant 20$	4
$20 < A \leqslant 40$	7
$40 < A \leqslant 60$	10
$60 < A \leqslant 80$	22
$80 < A \leqslant 100$	7

grade **C**

(a) Calculate an estimate for the mean area of these gardens.
(4 marks)

(b) Explain why the class interval that contains the median is $60\,\text{m}^2 < A \leqslant 80\,\text{m}^2$. **(2 marks)**

Scatter graphs

If the points on a scatter graph are almost in a straight line then the graph shows CORRELATION. The better the straight line, the stronger the correlation.

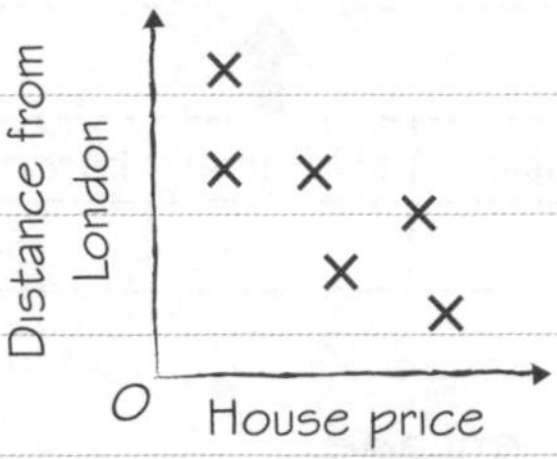

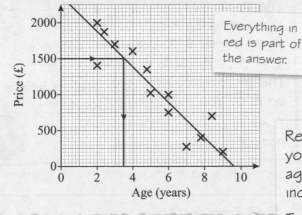

Negative correlation 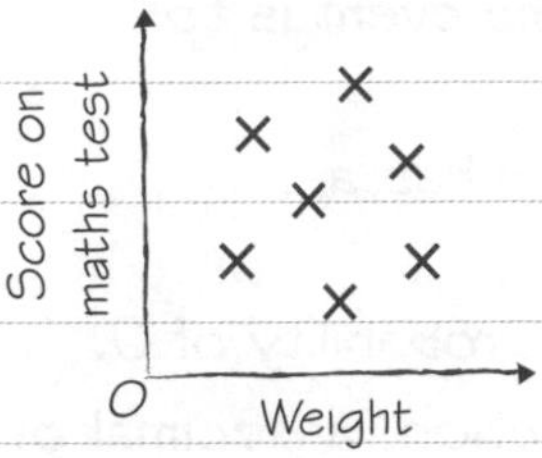　　No correlation 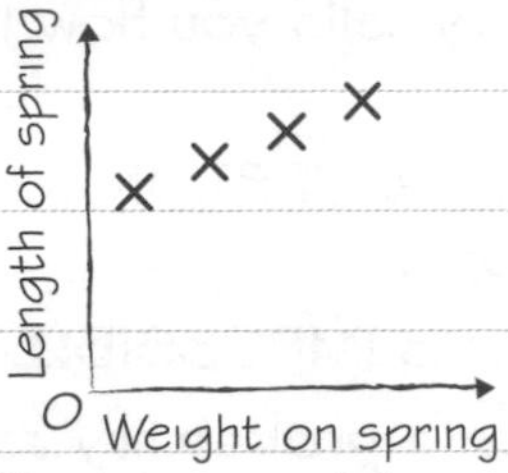　　Positive correlation

---

Worked example

grade **D**

A garage sells motorcycles.
The scatter graph gives information about the ages and prices of the motorcycles.

Everything in red is part of the answer.

(a) What type of correlation does the scatter graph show?

Negative

(b) Draw a line of best fit on the scatter graph.

Mae buys a motorcycle from the garage for £1500

(c) Use your line of best fit to estimate the age of the motorcycle.

3.5 years

Remember that the type of correlation tells you about the relationship between price and age. Negative correlation means that as the age increases the price decreases.

To predict the age of the motorcycle for part (c), read across from £1500 on the vertical axis to your line of best fit and then down to the horizontal axis. Draw the lines you use on your graph.

Now try this

grade **D**

edexcel

The scatter graph shows some information about six newborn baby apes.
For each baby ape, it shows the mother's leg length and the baby ape's birth weight.

A mother's leg length is 65 cm. Her baby ape's birth weight is 1.75 kg.

(a) Add this information to the scatter graph.
(1 mark)

(b) Describe the **correlation** between a mother's leg length and her baby ape's birth weight.
(1 mark)

(c) Draw a line of best fit on the diagram.
(1 mark)

A mother's leg length is 55 cm.

(d) Use your line of best fit to estimate the birth weight of her baby ape.
(1 mark)

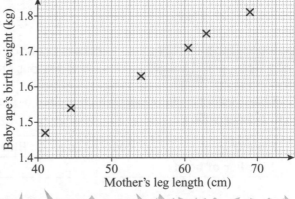

C
D
E
F
G

Probability 1

The probability that an event will happen is a value from 0 to 1.

The probability tells you how likely the event is to happen.

An event that is CERTAIN to happen has a probability of 1.

An event that is IMPOSSIBLE has a probability of 0.

You can write a probability as a fraction, a decimal or a percentage.

Impossible   Even chance   Certain

0   1

Fraction	Decimal	Percentage
$\frac{1}{2}$	0.5	50%

Worked example

grade **F**

(a) It is **very likely** that it will rain in Newcastle next October. Put a cross near 1 on the probability scale.

(a) On this probability scale, mark with a × the probability that it will rain in Newcastle next October.

0 |———————|————————×—| 1

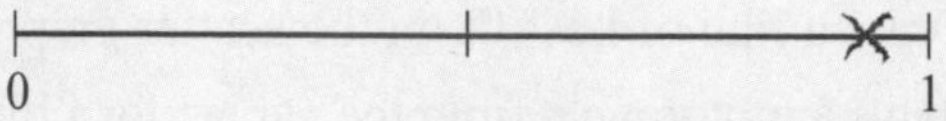

(b) Isobel says the probability she will be late for school is 7. Explain why Isobel is wrong.

Probabilities are numbers from 0 to 1.

Writing probabilities

The probability of rolling a 6 is $\frac{1}{6}$
You can write $P(6) = \frac{1}{6}$ 
There is one 6. There are six possible outcomes: 1, 2, 3, 4, 5, 6.

The probability of a coin landing heads up is $\frac{1}{2}$. You can write $P(Head) = \frac{1}{2}$

There is one head. There are two possible outcomes: head or tail.

Worked example

grade **F**

The diagram shows a fair 8-sided spinner.

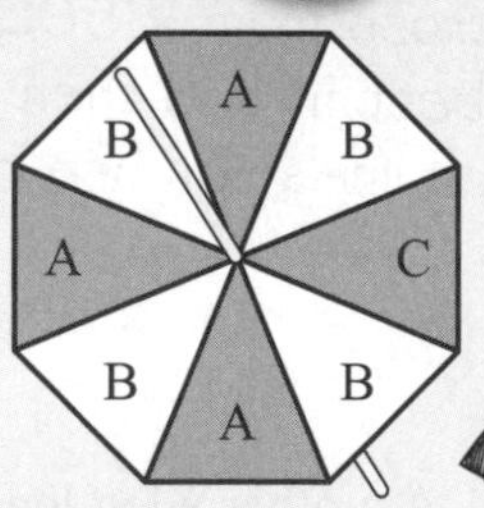

(a) Which letter is the spinner **least** likely to land on?

C

(b) Work out the probability that the spinner lands on the letter **A**.

$\frac{3}{8}$

Golden rule

$$Probability = \frac{number\ of\ successful\ outcomes}{total\ number\ of\ possible\ outcomes}$$

(a) There are 3 letter As. There are 4 letter Bs. There is 1 letter C. So C is the least likely result.

(b) There are 3 As. There are 8 possible outcomes
(A, B, C, B, A, B, A, B).

Now try this

Start by adding all the beads together to work out the total number of beads.

edexcel

A bag contains some beads which are red or green or blue or yellow.

Colour	Red	Green	Blue	Yellow
Number of beads	3	2	5	2

Samire takes a bead at random from the bag.

Write down the probability that she takes a blue bead. **(2 marks)**

grade **E**

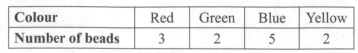

Probability 2

Probability that something will not happen

The probabilities (P) of all the different outcomes of an event add up to 1.

If you know the probability that something will happen, you can calculate the probability that it won't happen.

P(event doesn't happen) = 1 – P(event does happen)

The probability of rolling a 6 on a normal fair dice is $\frac{1}{6}$. So the probability of NOT rolling a 6 is $1 - \frac{1}{6} = \frac{5}{6}$

This spinner will definitely land on either red, yellow or green. So the probability of this happening is 1.

P(Red) + P(Yellow) + P(Green) = 1

Sample space diagrams

A SAMPLE SPACE DIAGRAM shows you all the possible outcomes of an event.
Here are all the possible outcomes when two coins are flipped.

First coin

	H	T
H	HH	TH
T	HT	TT

Second coin

There are four possible outcomes. TH means getting a tail on the first coin and a head on the second coin.

The probability of getting two tails when two coins are flipped is $\frac{1}{4}$ or 0.25. There are 4 possible outcomes and only 1 successful outcome (TT).

Worked example

grade **D**

Archie's sock drawer contains red, blue, black and grey socks.
The table shows the probabilities that a sock chosen at random will be red, blue or grey.

Colour	Red	Blue	Black	Grey
Probability	0.05	0.25		0.20

(a) Work out the probability that a sock chosen at random will be black.

P(Black) = 1 – (0.05 + 0.25 + 0.20)
        = 1 – 0.5 = 0.5

(b) Work out the probability that a sock chosen at random will be red or blue.

P(Red or Blue) = P(Red) + P(Blue)
            = 0.05 + 0.25 = 0.3

(a) The probabilities have to add up to 1. Use your calculator to add together the three probabilities you are given and then subtract the result from 1.

Check it!
0.05 + 0.25 + 0.5 + 0.20 = 1 ✓

(b) The question says 'red _or_ blue' so you need to _add_ together the probabilities for red and blue. You can leave your answer as a decimal number.

Now try this

grade **D**

Students can choose one of four snacks.
The table shows the probabilities that a student will choose burger or pizza or salad.

Snack	Burger	Pizza	Pasta	Salad
Probability	0.35	0.15		0.2

One student is chosen at random.
Work out the probability that the student
(a) did not choose salad   **(1 mark)**
(b) chose pasta   **(2 marks)**
(c) chose a burger or a salad.   **(2 marks)**

C
D
E
F
G

Probability 3

Relative frequency

You can calculate a probability from a two-way table or frequency table.

$$\text{Probability} = \frac{\text{frequency of outcome}}{\text{total frequency}}$$

This is called relative frequency.

Worked example — grade E

Shilpa asked her class whether they were left-handed or right-handed.
She recorded her results in a two-way table.

	Right-handed	Left-handed	Total
Boy	11	③	14
Girl	14	2	16
Total	25	5	㉚

A student is chosen at random from Shilpa's class. Work out the probability that the student is a left-handed boy.

$$\text{Probability} = \frac{3}{30} = \frac{1}{10}$$

There are 3 left-handed boys.
There is a total of 30 students.

Expectation

Probability helps you predict the outcome of an event.

If you flip a coin 100 times, you can expect to get heads about 50 times. You probably won't get heads exactly 50 times, but it's a good guess.

$$\text{Expected number of outcomes} = \text{number of trials} \times \text{probability}$$

Worked example — grade D

The probability of a biased coin landing heads up is 0.4

The coin is flipped 300 times.

Work out an estimate for the number of times the coin will land heads up.

$$300 \times 0.4 = 120$$

There are 300 trials and the probability of the coin landing heads up is 0.4

Fair or biased?

You can use expectation to help you decide if a dice or coin is FAIR.
These two coins have been flipped 50 times each.

Coin 1

Head	ҋ ҋ ҋ ҋ |||
Tail	ҋ ҋ ҋ ҋ ҋ ||

About the same number of heads and tails. This coin is probably fair.

Coin 2

Head	ҋ ҋ ||||
Tail	ҋ ҋ ҋ ҋ ҋ ҋ ҋ	

A lot more than the expected number of tails. This coin is probably BIASED.

Now try this

For a reminder about two-way tables have a look at page 90.

The two-way table gives some information about how 100 children travelled to school one day.

	Walk	Car	Other	Total
Boy	15		14	54
Girl		8	16	
Total	37			100

(a) Complete the two-way table. **(2 marks)**
One of the children is picked at random.
(b) Write down the probability that this child walked to school that day. **(2 marks)**
One of the girls is picked at random.
(c) Work out the probability that this girl did **not** walk to school that day. **(2 marks)**

grade D

Problem-solving practice

About half of the questions on your exam will need problem-solving skills.

These skills are sometimes called AO2 and AO3.

You can use a calculator on question 5, but practise the other questions without one. You might have to answer similar questions on your non-calculator paper.

For these questions you might need to:

- choose what method to use
- use the maths you've learnt in a new context
- plan your answer when solving harder problems
- show your working clearly and give reasons for your answers.

AO2
AO3

---

1 The table shows information about the numbers of Year 7 pupils absent from Keith's school last week.

grade
F

	Boys	Girls
Monday	8	10
Tuesday	11	9
Wednesday	12	12
Thursday	14	13
Friday	13	11

Keith wants to compare the data.

Draw a suitable diagram or chart.   (4 marks)

Bar charts p. 92

In this question **you** have to choose what type of diagram or chart to use. It is best to use a bar chart or a line graph.

TOP TIP

- Label BOTH axes correctly.
- Draw a key for boys and girls, or make sure it is clear which bars (or lines) are for boys and which are for girls.

---

2 Abi has five cards.
Each card has a number written on it.

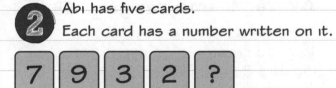

The mean of the five numbers is 6
One of the numbers is hidden.
Work out the hidden number.   (2 marks)

grade
D

Averages and range p. 95

You could try some different values for the hidden number and work out the mean each time. But you can save time by using the rule in the Top tip. There are five numbers and the mean is 6, so the sum of the numbers must be $5 \times 6 = 30$.

TOP TIP

This is a useful rule:

mean × number of data values
= sum of data values

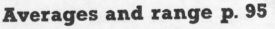

Problem-solving practice

3 The scatter graph shows the French and German marks of 15 students.

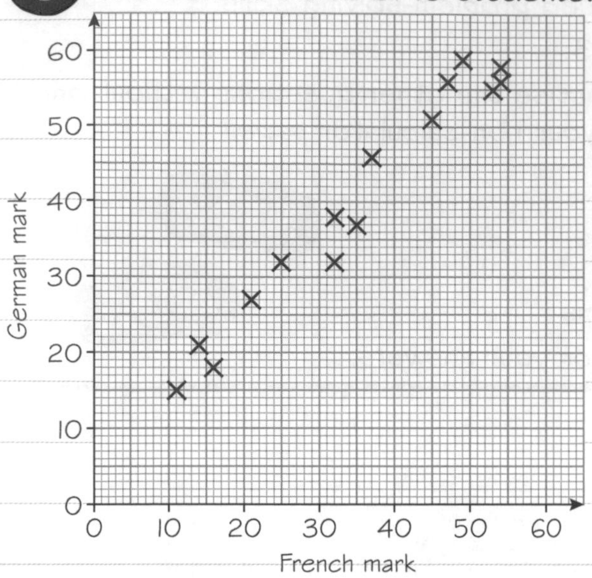
German mark
French mark
0
10
20
30
40
50
60
0
10
20
30
40
50
60

Jade's French mark was 42

Estimate Jade's German mark. (2 marks)

Scatter graphs p. 99

grade D

Start by drawing a line of best fit on the scatter graph. Next draw a line up from 42 on the horizontal axis to your line of best fit. Then draw a line across to the vertical axis to estimate Jade's German mark. Make sure you draw the lines you use on your graph to show your working.

TOP TIP

When you are reading information from a graph you should always give your answer to the nearest small square.

---

4 A teacher asked 30 students if they had a school lunch or a packed lunch or if they went home for lunch.

17 of the students were boys.

4 of the boys had a packed lunch.

7 girls had a school lunch.

3 of the 5 students who went home were boys.

Work out the number of students who had a packed lunch. (4 marks)

Two-way tables p. 90

grade D

You could get in a real mess with this question unless you draw a two-way table like this. Fill in all the given values and then complete the table.

	School lunch	Packed lunch	Home for lunch	Total
Boys				
Girls				
Total				30

TOP TIP

Start by filling in the total number of students.

---

5 *Some students in a class weighed themselves.

Here are their results.

Boys' weights in kg

70 65 45 52 63 72 63

Girls' weights in kg

65 45 47 61 44 67 55 56 63

Compare fully the weights of these students. (6 marks)

Averages and range p. 95

grade D

There are 6 **marks** for this question. To give a full answer you need to **compare** the data. So (1) calculate averages like the mean or median and a measure of spread like the range, and (2) write a sentence for each of these, comparing the boys and the girls.

TOP TIP

If a question has a * next to it, then there are marks available for QUALITY OF WRITTEN COMMUNICATION. This means you must show all your working and write your answer clearly with the correct units.

Formulae page

Area of trapezium $= \frac{1}{2}(a + b)h$

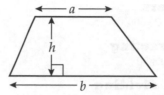

Volume of a prism $=$ area of cross section $\times$ length

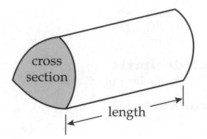

Answers

The number given to each topic refers to its page number.

NUMBER

1. Place value
(a) (i) fifty-four thousand, three hundred and twenty-seven
     (ii) 3 hundreds or 300
     (iii) 54 000
(b) 3091, 3100, 29 999, 35 687, 104 152

2. Rounding numbers
(a) 9800   (b) 23 600

3. Adding and subtracting
1. 38   2. £8.30

4. Multiplying and dividing
(a) 1610   (b) 46

5. Decimals and place value
1. (a) 9 hundredths or 0.09
   (b) (i) 0.067, 0.56, 0.6, 0.605, 0.65
       (ii) 0.07, 0.072, 0.7, 0.702, 0.72
2. (a) 195.2   (b) 1952

6. Operations on decimals
(a) £116.10   (b) (i) 28   (ii) £2.40

7. Estimating answers
1. 20   2. 14 000

8. Negative numbers
(a) (i) -8 (ii) $+15$ (b) -4 (c) $-5, -1, 0, 3, 8$

9. Squares, cubes and roots
1. (a) (i) 10   (ii) 2
   (b) (i) 49   (ii) 1   (iii) 12   (iv) 4
2. An example to show that the difference of two square numbers can be even: e.g. $3^2 - 1^2 = 9 - 1 = 8$

10. Factors, multiples and primes
1. (a) 6, 12, 30   (b) 3, 5   (c) 8 and 12
2. $2 \times 2 \times 2 \times 3 \times 7$ or $2^3 \times 3 \times 7$

11. HCF and LCM
(a) (i) $2 \times 2 \times 3 \times 5$ (ii) $2 \times 2 \times 2 \times 2 \times 2 \times 3$
(b) 12   (c) 480

12. Fractions
605

13. Simple fractions
(a) (i) $\frac{14}{15}$ (ii) $\frac{1}{6}$ (b) (i) $\frac{1}{2}$ (ii) $\frac{4}{5}$

14. Mixed numbers
(a) 10   (b) $1\frac{1}{4}$ (c) $3\frac{3}{4}$ (d) $1\frac{7}{8}$

15. Number and calculator skills
1. (a) 10   (b) 0   (c) 2   (d) 7
2. (a) 5.694 263 364   (b) 1.454 346 101

16. Percentages
(a) 656   (b) $144 < 200$ so Trudy incorrect
(c) 360   (d) 22%

17. Percentage change
1. £60.16   2. £7.60

18. Fractions, decimals and percentages
1. $\frac{2}{3}, \frac{4}{5}, 0.82, 85\%, \frac{7}{8}$ 2. £175

19. Ratio
1. 160 g sugar, 240 g butter, 400 g flour
2. £60

20–21. Problem-solving practice
1. No, e.g. $-3\,°C$ would be halfway between $-18\,°C$ and $12\,°C$.
2. £547.50 (by going to the theatre and sitting in the circle)
3. £54.45
4. 60 counters
5. Sportscentre Trainers (£45)

ALGEBRA

22. Collecting like terms
1. (a) (i) $4c$ (ii) $8g$ (iii) $4xy$
   (b) (i) $2a + 7b + 8$ (ii) $7r - 9t$
2. Jamie, because you can solve it to find x and an expression does not have an $=$ sign

23. Simplifying expressions
(a) p^4 (b) $10rp$ (c) $6x^2$
(d) $2f$ (e) $5d$ (f) 1

24. Indices
(a) p^2 (b) q^3 (c) t^{10} (d) m^6
(e) x^{12} (f) y^{14} (g) w^8

25. Expanding brackets
(a) (i) $12x - 12$ (ii) $3t^3 + 4t$ (iii) $3y^2 + 12y$
(b) (i) $17m - 9$ (ii) $-2y + 18$

26. Factorising
(a) (i) $3(p - 4)$ (ii) $t(t - 5)$
    (iii) $y(y + 1)$ (iv) $5(1 - 2v)$
(b) (i) $4(2a - 5)$ (ii) $3b(2b + 3)$
    (iii) $7c(2c - 3)$ (iv) $5x(2 + 5x)$

27. Sequences
(a) 17, 21   (b) 41
(c) No, because all the numbers are odd and 130 is even

28. nth term of a sequence
(a) $4n - 1$
(b) 318 is even and all the terms in the sequence are odd
(c) 3, 1, -1

29. Equations 1
(a) (i) $x = 6$ (ii) $t = 4$ (iii) $y = 10$ (iv) $m = 7$
(b) (i) $a = 7$ (ii) $b = -1$ (iii) $c = 9$

30. Equations 2
1. (a) $x = \frac{1}{2}$ (b) $x = \frac{2}{5}$
2. $y = 3\frac{1}{2}$

31. Writing equations
(a) $x + 2$ (b) $4x + 14$ (c) 1.5

32. Trial and improvement
1. 2.7   2. 3.6

33. Inequalities
(a) $x > -1$
(b) (i)

   (ii) $-3, -2, -1, 0, 1$

34. Solving inequalities
1. (a) $x \leqslant 2$ (b) $y > -3$
2. (a) $x \geqslant -3.5$ (b) -3

35. Substitution
1. (a) 1   (b) 27   (c) -5
2. Bryani, because need to square first before $\times$ 4

36. Formulae
(a) £22   (b) 15

37. Writing formulae
(a) $P = 6x + 4$ (b) 34 cm

38. Rearranging formulae
1. 300 miles
2. (a) $t = \dfrac{v - u}{5}$ (b) $q = \dfrac{d}{a} + 3$ (c) $a = 4(s - 2u)$

39. Coordinates
(a) $(1, 4)$ (b) R plotted at $(5, -2)$
(c) $(1, 2)$ (d) $(3, 2)$

40. Straight-line graphs 1
(a) 2   (b) $-\dfrac{6}{5}$

41. Straight-line graphs 2
(a) graph of $y = 3x + 1$ drawn   (b) 3

42. Real-life graphs
(a) $45   (b) €60
(c) Cheaper in America because €100 = $150 > $120

43. Distance–time graphs
(a) 30 km   (b) 45 minutes   (c) 60 km/h

44. Interpreting graphs
(B), D, C, A, F, E

45. Quadratic graphs
(a) $5, -4, -3$ (b)

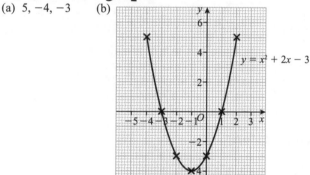

46. Using quadratic graphs
(a) $3, -6, -5$ (b)

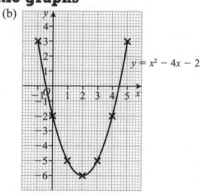

(c) 0.3 and 3.7   (d) -0.45 and 4.45

47–48. Problem-solving practice
1. (a) 14 dots
   (b) No, all the patterns have an even number of dots and 21 is an odd number.
2. £92
3. 1.4 kg
4. $t = 4\frac{1}{3}$
5.

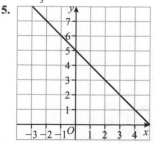

GEOMETRY AND MEASURES

49. Measuring and drawing angles
(a) $28° \pm 2°$ (b) $307° \pm 2°$

50. Angles 1
(a) R marked at top left or bottom left
(b) (i) acute   (ii) reflex
(c) 325° (angles at a point add up to 360°)

51. Angles 2
angle $AQR = 72°$ (angles on a straight line add up to 180°)
angle AQR = angle CRS (corresponding angles are equal)
So $x = 72°$

52. Solving angle problems
1. (a) 30° (base angles of isosceles triangle are equal, angles in a triangle add up to 180°)
   (b) 48° (angles on a straight line add up to 180°, angles in a quadrilateral add up to 360°)
2. (a) (i) 72°   (ii) alternate angles are equal
   (b) 58°

53. Angles in polygons
(a) 9   (b) 15   (c) 144°

54. Measuring lines
(a) 1.5 metres to 2 metres   (b) 4.5 metres to 6 metres

55. Bearings

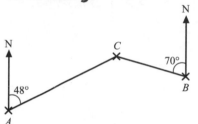

56. Scale drawings and maps
(a) 078°   (b)

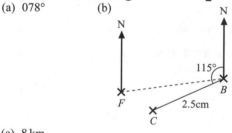

(c) 8 km

57. Symmetry
(a) B and D   (b) (i) A   (ii) 3

58. 2-D shapes
(a) square   (b) kite drawn on a grid   (c) trapezium

59. Congruent shapes
e.g.

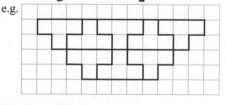

60. Reading scales
(a) 48 marked   (b) 6.37 marked

61. Time and timetables
(a) (i) 15 minutes   (ii) 09:45   (b) 80 minutes

62. Metric units
capacity of container $= 19.5 \times 1000 = 19\,500$ ml
number of full cups possible $= \dfrac{19\,500}{210} = 92.8571\ldots$
So at most 92 cups can be filled completely

63. Measures

1. 70 mph (Great Britain) is lower because 70 mph = 112 km/h or 120 km/h = 75 mph

2. Jane: 1 hour 36 minutes   Mattie: 1 hour 40 minutes

64. Speed

(a) 40 mph   (b) 16 minutes

65. Perimeter and area

(a) 18 cm   (b) 14 cm^2

66. Using area formulae

12 m^2

67. Solving area problems

20 sheep (area = 102 m^2)

68. Circles

(a) 2.04 m   (b) 489.7

69. Area of a circle

30.9 cm^2

70. 3-D shapes

1. (a) sphere   (b) cylinder   (c) tetrahedron
2. Lia, as 12 + 2 = 6 + 8

71. Plan and elevation

(a) 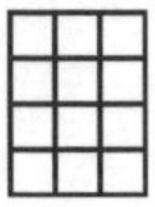   (b)

72. Volume

100

73. Prisms

(a) 42 cm^3
(b) $\frac{1}{2}(4 \times 3) + \frac{1}{2}(4 \times 3) + (7 \times 5) + (7 \times 3) + (7 \times 4)$
    $= 6 + 6 + 35 + 21 + 28 = 96$ cm^2

74. Cylinders

(a) 2260 cm^3 (2261.9)   (b) 867 cm^2

75. Units of area and volume

1. (a) 25 000 cm^2 (b) 85.6 cm^2
    (c) 7000 mm^3 (d) 0.391 m^3
2. 30 000 litres

76. Constructions 1

1.

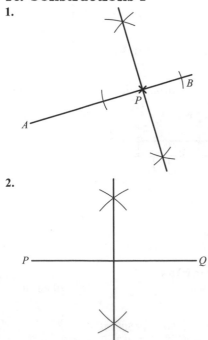

2.

77. Constructions 2

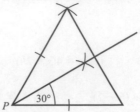

78. Loci

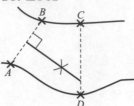

79. Translations

1.

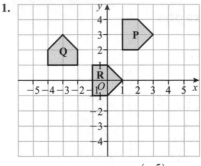

2. Translation by the vector $\begin{pmatrix} -5 \\ -1 \end{pmatrix}$

80. Reflections

1.

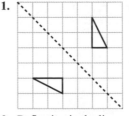

2. Reflection in the line $y = x$

81. Rotations

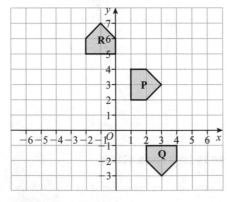

82. Enlargements

1.

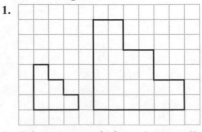

2. Enlargement, scale factor 3, centre (0, 0)

108

83. Combining transformations

(a)
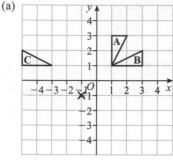

(b) Reflection in the line $x = -1$

84. Similar shapes

(a) 2   (b) 120°

85. Pythagoras' theorem

84 cm²

86. Line segments

1. 10 units
2. 8.94 units

87–88. Problem-solving practice

1. e.g.
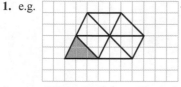

2. e.g. $\angle DEB = 105°$ (opposite angles in parallelogram are equal)
   $x = 75°$ (angles on a straight line add up to 180°)
3. 13.8 km

4. Rotation of 90° clockwise about $(0, -4)$
5. 8 flowerpots

STATISTICS AND PROBABILITY

89. Collecting data

1. (a) e.g. overlapping regions, no time frame
   (b) How many hours, to the nearest hour, did you use your mobile phone last week?

   ☐ 0–1   ☐ 2–3   ☐ 4–5   ☐ more than 5

90. Two-way tables

1.

	Year group			Total
	9	**10**	**11**	
Boys	149	133	125	407
Girls	154	123	147	424
Total	303	256	**272**	831

2. (a)

	London	York	Total
Boys	23	14	37
Girls	19	24	43
Total	42	38	80

(b) 24

91. Pictograms

(a) 12   (b) 6   (c) 2 circles and $1\frac{1}{4}$ circles

92. Bar charts

(a) 85 minutes   (b) Wednesday and Friday
(c) 320 minutes (or equivalent)

93. Frequency polygons

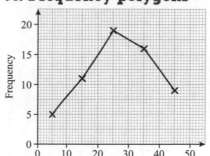

94. Pie charts

1. (a) 16   (b) 72
2.
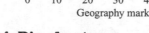

95. Averages and range

(a) 3
(b) 4
(c) 5.1
(d) No as boys' range (7) is the same as girls' range (7)

96. Stem and leaf diagrams

14	7 7 7 8
15	1 2 3 5 8 8
16	4 5 6 6 7
17	1 2 8
18	9 9

Key: 17 | 8 means 178 (mm)

97. Averages from tables 1

1. (a) Mode = 7 as 12 houses had 7 rooms
   (b) 237
2. (a) 12   (b) 3   (c) 13.3

98. Averages from tables 2

(a) 58.4 m²
(b) The 25th and 26th values are in the group $60 < A \leq 80$

99. Scatter graphs

(a) point (65, 1.75) plotted   (b) positive
(c) line of best fit drawn   (d) ~1.65 kg

100. Probability 1

$\frac{5}{12}$

101. Probability 2

(a) 0.8   (b) 0.3   (c) 0.55

102. Probability 3

(a)

	Walk	Car	Other	Total
Boy	15	**25**	14	54
Girl	**22**	8	16	**46**
Total	37	33	**30**	100

(b) $\frac{37}{100}$ or equivalent

(c) $\frac{24}{46}$ or equivalent

103–104. Problem-solving practice

1. e.g.

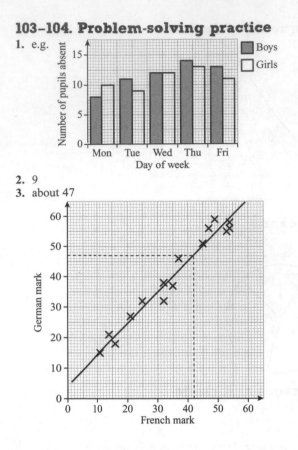

2. 9

3. about 47

4. 4

5. e.g.

	Mean	Median	Range
Boys	61.4 kg	63 kg	27 kg
Girls	55.9 kg	56 kg	23 kg

The boys had a higher mean and median than the girls, so they were on average heavier.
The boys had a higher range than the girls, so their weights were more spread out.

Published by Pearson Education Limited, a company incorporated in England and Wales, having its registered office at Edinburgh Gate, Harlow, Essex, CM20 2JE. Registered company number: 872828

www.pearsonschoolsandfecolleges.co.uk

Text © Keith Pledger, Harry Smith and Pearson Education Limited 2011
Edited by Fiona McDonald and Laurice Suess
Typeset by Tech-Set Ltd, Gateshead
Original illustrations © Pearson Education Limited 2011

The rights of Keith Pledger and Harry Smith to be identified as authors of this work have been asserted by them in accordance with the Copyright, Designs and Patents Act 1988.

First published 2011

17 16 15 14
14 13 12 11

British Library Cataloguing in Publication Data
A catalogue record for this book is available from the British Library

ISBN 978 1 44690 017 8

Printed in Slovakia by Neografia

Disclaimer
This material has been published on behalf of Edexcel and offers high-quality support for the delivery of Edexcel qualifications.
This does not mean that the material is essential to achieve any Edexcel qualification, nor does it mean that it is the only suitable material available to support any Edexcel qualification. Material from this publication will not be used verbatim in any examination or assessment set by Edexcel. Any resource lists produced by Edexcel shall include this and other appropriate resources.

Copies of official specifications for all Edexcel qualifications may be found on the Edexcel website: www.edexcel.com